NOTICE TO THE READER

Publisher does not warrant or guarantee, any of the products described herein or perform any independent analysis in connection with any of the product information contained herein. Publisher does not assume, and expressly disclaims, any obligation to obtain and include information other than that provided to it by the manufacturer.

The reader is expressly warned to consider and adopt all safety precautions that might be indicated by the activities described herein and to avoid all potential hazards. By following the instructions contained herein, the reader willingly assumes all risks in connection with such instructions.

The publisher makes no representations or warranties of any kind, including but not limited to, the warranties of fitness for particular purpose or merchantability, nor are any such representations implied with respect to the material set forth herein, and the publisher takes no responsibility with respect to such material. The publisher shall not be liable for any special, consequential or exemplary damages resulting, in whole or in part, from the readers' use of, or reliance upon, this material.

DELMAR PUBLISHERS INC.
2 COMPUTER DRIVE, WEST — BOX 15-015
ALBANY, NEW YORK 12212

Preface

Mathematics is involved in every phase of carpentry work including blueprint reading, measurement of materials, and layouts. *Mathematics for Carpenters* was developed to meet the need for a comprehensive text which combines mathematical concepts with information on the skills and techniques of carpentry.

The principles presented in the text range from a basic study of whole numbers to the fundamentals of trigonometry. These principles are introduced in an easy-to-understand format and are combined with specific information on the methods, practices, and tools of carpentry. Metric information, reviewed by the National Bureau of Standards for accuracy and correct usage, is included. The units are organized in a logical sequence with each unit based on the preceding unit. Each unit is accompanied by solved problems which explain the mathematical concepts in a step-by-step manner.

Performance objectives at the beginning of each unit state what the student is expected to accomplish before proceeding to the next unit. Review sections involving both mathematical and carpentry concepts are presented at the end of each unit. Some problems are designed to familiarize the carpentry student with other construction trades, such as masonry and plumbing. Many of the review questions are related to accompanying illustrations. The Appendix of the text includes a Glossary of carpentry and mathematical terms, measurement conversion tables, and a table of standard blueprint symbols which the carpenter must know to read building plans.

A four-year apprenticeship program is generally considered the best way to learn carpentry. The apprentice usually attends 144 hours of related classroom instruction each year and is compensated throughout this time for his work experience. *Mathematics for Carpenters* was designed with the union-based apprentice and other carpentry students in mind, for whom a basic mathematical ability is essential.

The author, Robert Bradford, is presently Associate Professor of Industrial Arts Education and Industrial Technology at Grambling State University, Grambling, Louisiana. He has been employed as a builder, supervisor, consultant, and inspector of residential and commercial construction. Mr. Bradford is a member of the American Vocational Association and the Association of Higher Education.

Other texts in the Delmar Trade-Related Mathematics series include:

> *Mathematics for Plumbers and Pipe Fitters*
> *Mathematics for Machine Technology*
> *Mathematics of the Shop*
> *Mathematics for Sheet Metal Fabrication*

Texts in the Delmar Carpentry/Woodworking series include:

> *Units in Woodworking*
> *The Use of Hand Woodworking Tools*
> *The Use of Portable Power Tools*
> *Framing, Sheathing, and Insulation*
> *Carpentry*
> *Understanding Construction Drawings*
> *Concrete Form Construction*
> *Blueprint Reading and Sketching for Carpenters — Residential*
> *Practical Problems in Mathematics for Carpenters*

Mathematics for Carpenters

*This book is dedicated
to my wife,
Ruth*

Mathematics for Carpenters

Robert Bradford

15 14 13 12 11

LIBRARY OF CONGRESS CATALOG CARD NUMBER: 75-19525
ISBN: 0-8273-1116-8

Printed in the United States of America
Published simultaneously in Canada
by Nelson Canada,
A Division of International Thomson Limited

Contents

The author and editorial staff at Delmar Publishers are interested in continually improving the quality of instructional material. The reader is invited to submit constructive criticism and questions. Responses will be reviewed jointly by the author and source editor. Send comments to:

Editorial Department
Delmar Publishers Inc.
2 Computer Drive-West Box 15-015
Albany, New York 12212

CAREER PROFILE: THE ESTIMATOR

Job Description

Contractors hire estimators to carefully figure the cost of material, labor, equipment, and subcontract work needed for a construction job. They are usually subcontracted or salaried, but may be self-employed.

Estimators are specialists in reading blueprints and interpreting building specifications. They must be aware of prices of materials and labor. Generally, they must have the ability to accurately estimate the complete cost of construction. Estimators spend considerable time at the building site, but most of their work is done indoors.

Qualifications

Estimators are required to have a solid background in mathematics, industrial materials, blueprint reading, and interpreting specifications. Students may take related courses in science and drafting.

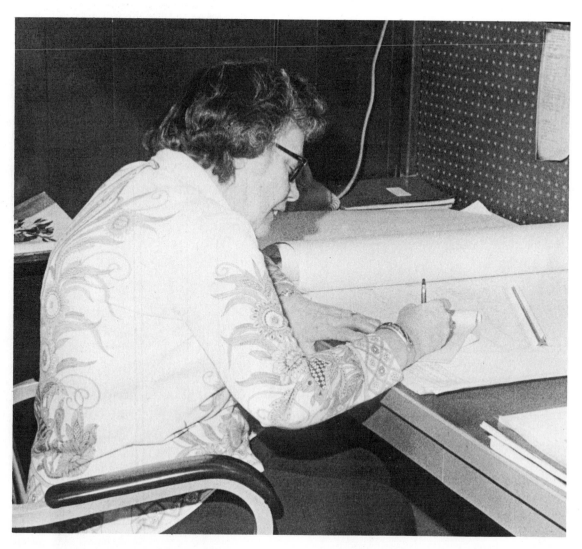

Section 1
Whole Numbers

Unit 1
Reading and Writing Whole Numbers

OBJECTIVES

After studying this unit the student should be able to

- read and write the mathematical signs and abbreviations used in carpentry.
- read and write small and large whole numbers.
- apply the use of whole numbers in solving carpentry problems.

Numbers play an important part in the everyday activity on a construction site. Recognizing signs and abbreviations as well as reading and writing whole numbers is essential to solving problems in carpentry.

SIGNS AND ABBREVIATIONS

Figure 1-1 illustrates the four arithmetic processes commonly used in construction and their representative signs.

Figure 1-2 lists terms and abbreviations often used in construction.

ADDITION
5 + 2 = 7
5 plus 2 = 7
5 and 2 = 7
SUBTRACTION
5 − 2 = 3
5 minus 2 = 3
5 less 2 = 3
MULTIPLICATION
5 x 2 = 10
5 · 2 = 10
(5) (2) = 10
[5] x [2] = 10
5 times 2 = 10
DIVISION
10 ÷ 5 = 2
$5\overline{)10}^{\,2}$
$5\underline{)10}^{\,2}$

Fig. 1-1 Processes and Signs

Foot or feet:	ft., '
Inches	in., "
Square feet:	sq. ft.
Square inches	sq. in.
Square yards	sq. yd.
Cubic feet	cu. ft.
Cubic inches	cu. in.
Cubic yards	cu. yd.
Board foot or board feet	bd. ft.
Pounds	lb.
Degrees	deg., °
Minutes	min., '
Seconds	sec., "

Fig. 1-2 Abbreviations of Measurements

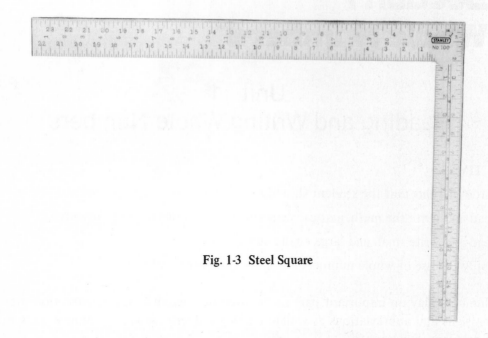

Fig. 1-3 Steel Square

READING AND WRITING SMALL WHOLE NUMBERS

Small numbers may be considered as whole numbers containing one or two digits. The Arabic numerals or notations (1, 2, 3, 4, 5, 6, 7, 8, 9, and 0) are called *digits* or *figures*. The *notation* of a number is its symbol. These ten figures can be used in any combination to express all whole numbers. The zero (0) has no value by itself.

Numbers are obtained by counting or measuring items. Some of the numbers found on the *steel square* are small whole numbers, figure 1-3. The steel square (also called *framing square*) is used as a layout tool in all construction areas: carpentry, plumbing, masonry, and sheet metal work.

The zigzag rule, figure 1-4, is usually 72″, or 6′, long; there are 12″ in each foot. Therefore, measurements can be expressed as whole inches or whole feet. The zigzag rule is used in layouts to measure relatively short distances. *Layout* is the process of accurately marking material with a given measurement.

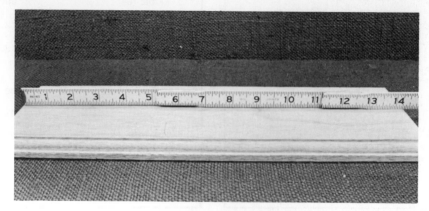

Fig. 1-4 Zigzag Rule

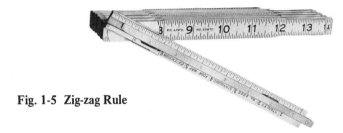

Fig. 1-5 Zig-zag Rule

Illustration. A board is measured as 14 inches long. In what ways can this measurement be expressed?

Solution. It can be written as fourteen inches, 14 inches, 14 in., or 14″.

Explanation. • Words or symbols may be used to express the number. The correct abbreviation is annexed to the figure. To *annex* is to add a figure to the end of a number.

Illustration. A wall measures 4 feet high with a zig-zag rule. Express the measurement in several ways.

Note: Zig-zag rules have graduations in feet and inches, figure 1-5.

Solution. Four feet can be written as four feet, 4 feet, 4 ft., or 4′.

Explanation. • The numerical measurement is written as words or figures and the correct abbreviation is annexed.

Illustration. Express the measurement of a pipe 21 inches long.

Solution. It can be expressed as twenty-one inches, 21 inches, 21 in., or 21″.

Explanation. This case is similar to the two former illustrations except that this illustration involves reading and writing a two-digit number. Compound words used to express numbers less than 100 are hyphenated. Therefore, it is necessary to use a hyphen (-) between the words twenty and one.

READING AND WRITING LARGE WHOLE NUMBERS

Large numbers can be defined as whole numbers containing three or more digits. Making perimeter measurements, area measurements, volume measurements, and load calculations are some of the instances in which a carpenter must be able to read and write large whole numbers.

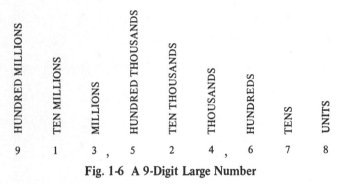

HUNDRED MILLIONS	TEN MILLIONS	MILLIONS	HUNDRED THOUSANDS	TEN THOUSANDS	THOUSANDS	HUNDREDS	TENS	UNITS
9	1	3 ,	5	2	4 ,	6	7	8

Fig. 1-6 A 9-Digit Large Number

The numbers in figure 1-6 are arranged in groups of three figures separated by commas. This method makes it easier to read and write large numbers; 913,524,678 is easier to read than 913524678. The number shown is read nine hundred thirteen million five hundred twenty-four thousand six hundred seventy-eight.

Illustration. A house measures 2,176 square feet. Express this figure in words.

Solution. Two thousand one hundred seventy-six square feet.

Explanation. • The number is expressed in words which name the symbols used. The symbols indicate the position of the number. The number could be expressed in a shortened version: twenty-one hundred seventy-six square feet.

Illustration. The allowable fiber stress for a solid wood girder is one thousand six hundred pounds per square inch, figure 1-7. Express the weight in figures.

Solution. 1,600 lbs. per sq. in.

Explanation. • The first 0 on the right is units; the second digit, 0, is tens; the third digit, 6, is hundreds; the fourth digit, 1, is thousands. • Starting at the left and proceeding to the right, each group is read as if it were alone. • The abbreviations are then annexed. The word *and* should not be used when reading whole numbers.

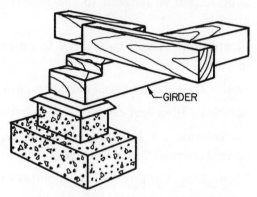

Fig. 1-7 Fiber Stress on a Wood Girder

A complete list of signs and abbreviations used in construction work is included in the Appendix.

APPLICATION

1. Write the signs for the following processes.

   a. addition
   b. subtraction
   c. multiplication
   d. division

2. Write the abbreviations for the following words.

   a. inch or inches
   b. foot or feet
   c. square
   d. pound
   e. yard
   f. cubic
   g. square feet
   h. cubic feet
   i. cubic yard
   j. board foot

3. Write the words which represent each number listed below.

   a. 1
   b. 2
   c. 3
   d. 4
   e. 5
   f. 6
   g. 7
   h. 8
   i. 9
   j. 10
   k. 11
   l. 12
   m. 13
   n. 14
   o. 15
   p. 16
   q. 17
   r. 18
   s. 19
   t. 20

4. The *try square*, figure 1-8, is frequently used in layout procedures for carpentry and sheet metal work. Write the correct words for each numeral on the illustrated try square.

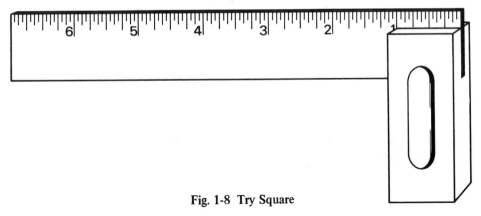

Fig. 1-8 Try Square

5. Use numbers to express the following words.

a. twenty-five	g. seventy-two	m. eighty-six	s. ninety-seven
b. sixteen	h. eighty-eight	n. ninety-eight	t. seventy-four
c. nineteen	i. seventy-five	o. thirty	u. ninety
d. twenty-three	j. sixty-six	p. thirty-three	v. thirty-eight
e. ninety-nine	k. sixty-seven	q. forty-seven	
f. eighty-five	l. sixty-four	r. fifty-eight	

6. Express each number in words.

a. 45,005	g. 432,016	m. 19,877,552
b. 96,000	h. 55	n. 220,666,751
c. 211,336	i. 913,845	o. 966,157,001
d. 610,234	j. 2,904,000	p. 50
e. 28	k. 4,423,680	q. 678,855,212
f. 798,285	l. 7,533,825	r. 785,751,616

7. Express each of the following words in numbers.

a. Fifty-seven million fifty-seven thousand ten

b. Seven thousand eight hundred

c. Thirteen million two hundred fifteen

d. Three thousand fifty

e. One hundred thousand nine hundred eighty-nine

8. An *extension rule,* which extends to 10 feet, is shown in figure 1-9. Make a two-column chart. On the left side write the figures which appear on the rule and on the right side, the corresponding words. Annex the correct abbreviation.

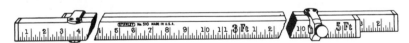

Fig. 1-9 Extension Rule

9. Place a comma where needed in each of the following numbers.

a. 5809 sq. in.	f. 950670 sq. yd.	k. 9100000 cu. yd.
b. 2579 sq. in.	g. 867059 sq. yd.	l. 8105600 cu. yd.
c. 9349 sq. ft.	h. 765805 sq. yd.	m. 50687958 cu. ft.
d. 9506 sq. ft.	i. 658999 cu. in.	n. 411224060121 lb.
e. 10700 sq. ft.	j. 100000 cu. in.	o. 521232760326 lb.

10. Insert hyphens where needed in each of the following numbers.

a. Five thousand two hundred thirty five

b. Two hundred four thousand sixty eight

c. Eighty two thousand

d. Fifty five hundred

e. One hundred forty two million two thousand twenty

Unit 2
Addition of Whole Numbers

OBJECTIVES

After studying this unit, the student should be able to

- add whole numbers.
- solve problems in carpentry which require addition.
- check addition for correctness.

The process of combining two or more numbers to make one number of equal value is called *addition*. The numbers which are added are known as *addends* and the result is called the *sum*. Addition is the most commonly performed mathematical process in construction. Addends must be expressed in the same units of measurement. For example, feet and pounds cannot be combined by simply adding the numbers. When like units are added, the answer should contain the name of the unit.

Illustration. A carpenter saws two pieces of lumber for batter boards, figure 2-1. *Batter boards* are pieces of lumber set up around corners and partitions to locate building lines during foundation construction. One piece (pc.) is 6′ and the other is 5′ long. What is the total length of the two pieces?

Solution. 6′ + 5′ = 11′

Explanation. • Both pieces of lumber are measured in feet; therefore, they can be added without unit conversion. • The process involves combining 6′ and 5′ and annexing the symbol to represent feet. The sum is 11′.

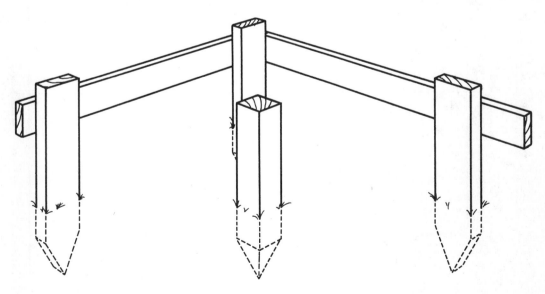

Fig. 2-1 Batter Boards

VERTICAL ADDITION

When numbers are arranged for addition, the addends are usually placed in vertical columns:

addend	6	5
+ addend	+ 5	+ 6
sum	11	11

The carpenter should be able to add two or more numbers, each containing two or more digits. The units (tens, hundreds, and thousands) of each number must be lined up in their respective columns. Avoid error in addition by keeping the columns straight. Add the digits in each column and use the method of carrying to obtain the sum.

Illustration. Find the total number of feet around the building line layout in figure 2-2.

Solution.

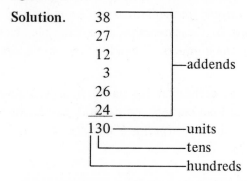

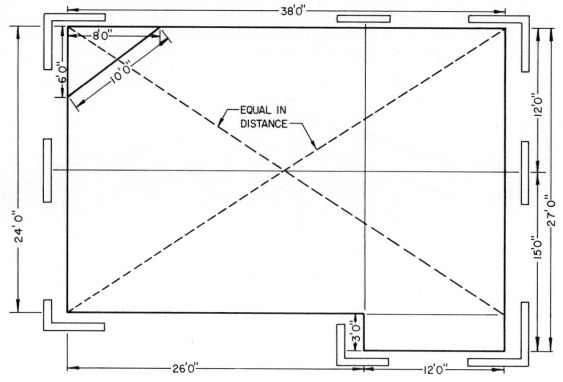

Fig. 2-2 Building Lines

Explanation. • The sum of digits in the units column is 30. Write 0 in the units column and mentally carry 3 to the tens column. • The sum of the digits in the tens column, including the 3 that is carried, is 13. • Write the 13 with 3 in the tens column and 1 in the hundreds column. There is 130′ around the building.

CHECKING VERTICAL ADDITION

Checking addition for correctness is especially important in carpentry work. A method of checking addition is to add upward and then add downward.

Illustration. Check for the correct answer by adding up and then down.

Solution.

```
     1105  Sum
       80
       80
      176
      121
      216
      216
      216
     1105  Sum
```

Explanation. • Add the units, tens, and hundreds columns from bottom to top. • Write the sum at the top of the problem. • Add the columns in the same order from top to bottom. • Write the sum at the bottom and compare the answers.

Adding upward, and then downward, changes the order in which the figures are added. If an error is made in the addition, the sums will not be the same. If the two sums are the same, the answer is probably correct.

APPLICATION

1. Add the numbers in each of the following.

a. 1 2 +3	e. 8 9 +3	i. 7 9 +6	m. 2 9 +7	q. 7 5 +3
b. 2 3 +4	f. 7 8 +4	j. 9 7 +5	n. 5 3 +8	r. 8 8 +9
c. 4 5 +6	g. 4 7 +6	k. 9 8 +1	o. 6 4 +1	
d. 6 7 +8	h. 8 5 +9	l. 6 4 +2	p. 4 2 +9	

2. The architectural drawings shown below indicate that the distance from the floor to the top of a window is 7′, and from that point to the ceiling is 1′. What is the height of the wall? What is the total height of the elevation?

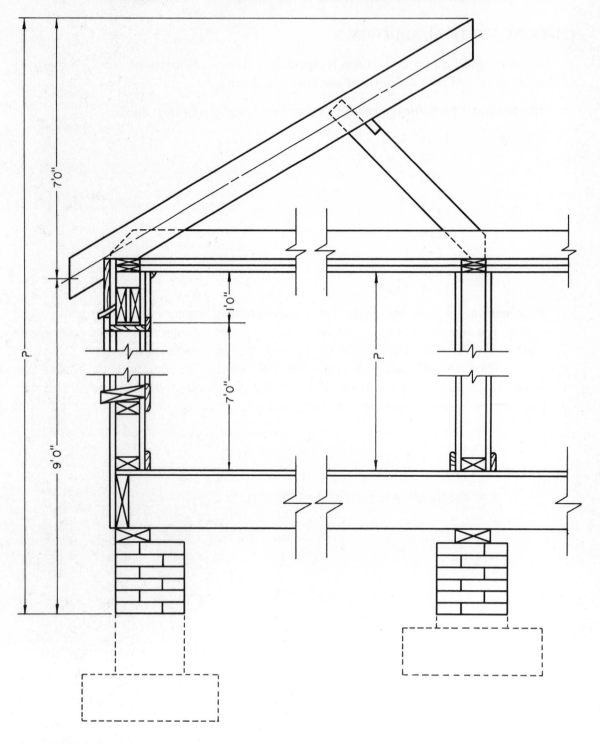

3. Add and check the numbers in each of the following.

a.	11	c.	18	e.	14	g.	17
	15		19		3		6
	+ 19		+ 19		+ 5		+ 9

b.	16	d.	17	f.	19	h.	12
	19		16		7		16
	+ 18		+ 13		+ 6		+ 7

4. The electrical fuses used in the average-sized home are usually rated at 15 or 20 amperes. What is the total number of amperes in the following fuses: 15, 20, 15, 20, 20, 15, 15, and 20?

Note: A *fuse* is an electrical device made with a thin strip of metal which melts and interrupts excessive currents. In this way, fuses protect against overloaded circuits. An *ampere* is the amount of current passing a fixed point in a conductor each second.

5. Add and check for correctness.

5	10	12334	23540
55	133	1956	2376
258	1313	563	269
+ 1563	+ 21250	+ 7	+ 5

6. Measurements on the floor plan in the figure below show the width of the structure in sections. Find the total width of the structure.

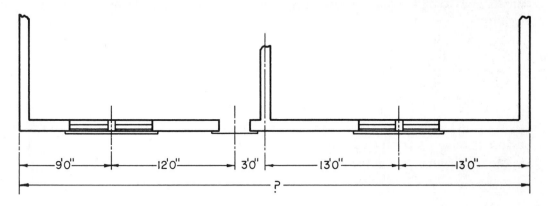

7. Labor costs on a house include skilled labor, $12,800; apprentice labor, $7,500; and unclassified labor, $350. What is the total labor cost?

8. Material costs on construction of a house include foundation, $2256; main frame, $4026; roof frame, $1286; exterior covering and finish, $2800; interior finish, $3987. What is the total material cost? Check for correctness.

Unit 3
Subtraction of Whole Numbers

OBJECTIVES

After studying this unit the student should be able to

- find the difference between two numbers.
- subtract numbers.
- solve problems in carpentry requiring combined addition and subtraction.

Subtraction is the process of determining the difference between two numbers. In the addition process, two numbers are combined; in subtraction, a smaller number is taken from a larger number. The larger number is the *minuend* and the smaller number is the *subtrahend.* The answer obtained is called the *difference* or *remainder.*

SUBTRACTING TWO NUMBERS

When numbers contain two or more digits, it is usually convenient to use the vertical method of subtraction. The procedure is similar to vertical addition in that there are separate columns for the tens, hundreds, and thousands units. In subtraction, however, the larger number (minuend) must be placed above the smaller number (subtrahend).

Illustration. A carpenter has a 50′ steel tape of which 20′ is accidently destroyed. How many feet are left? *Note:* The steel tape, figure 3-1, is used for measurements and layouts involving long distances. They are available in lengths from 50′ to 300′.

Solution. 50′ minuend
       − 20′ subtrahend
        30′ remainder

Explanation. • The minuend (50) is placed above the subtrahend (20). • Begin at the right in the units column and take 0 units from 0 units. Write 0 in the units column below the line. • Subtract 2 tens from 5 tens in the tens column, leaving 3 tens. Write in the tens column below the line. Therefore, 30′ remain on the steel tape.

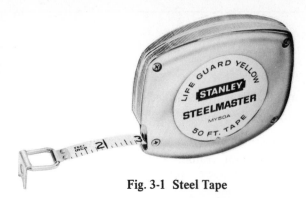

Fig. 3-1 Steel Tape

Sometimes it is necessary to borrow when subtracting one number from another. Borrowing is necessary when the number in a column in the subtrahend is larger than the number in the corresponding column in the minuend.

Illustration. A builder orders 74′ of lumber. Of this amount, 39′ are delivered. How many feet are not delivered?

Solution.
$$\begin{array}{r} 74' \\ -\ 39' \\ \hline 35' \end{array}$$

Explanation. • Since 9 units cannot be subtracted from 4 units, borrow 1 ten from 7 tens. • One ten is equal to 10 units: 10 units + 4 units = 14 units. • Subtract 9 units from 14 units which leaves 5 units. • Borrowing 1 ten from 7 tens changes the 7 tens to 6. • Subtract the 3 in the tens column from the 6. This leaves 3 tens. The remainder is 35 feet.

Illustration. A carpenter has 827 sq. ft. of flooring to install in a playroom, figure 3-2. If 538 sq. ft. are installed the first day, how many feet are left to install?

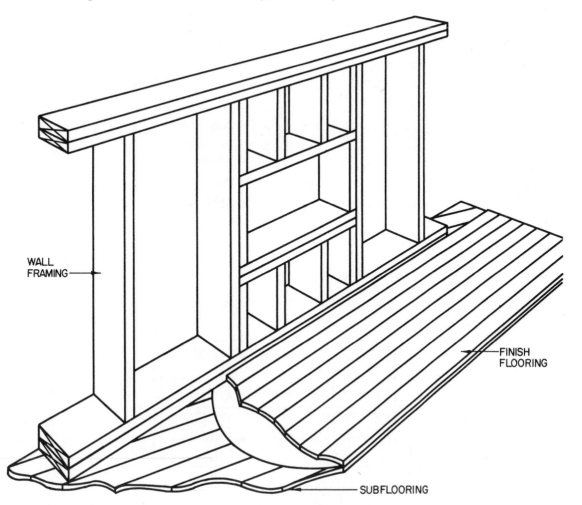

WALL FRAMING

FINISH FLOORING

SUBFLOORING

Fig. 3-2 **Wall Framing, Subflooring, and Finish Flooring.**

Solution.

$$827 \text{ sq. ft.}$$
$$- \underline{538} \text{ sq. ft.}$$
$$289 \text{ sq. ft.}$$

Explanation. • In this case, 8 cannot be taken from 7, so 1 is taken from the tens place in the minuend and added to the 7. • Since one ten equals 10 units, 10 + 7 = 17. • Subtract 8 from 17, leaving 9.

After 1 ten is taken from the tens place, 1 ten is left in the second, or tens, column. • Since 3 cannot be taken from 1, take 1 hundred from the hundreds place and add it to the 1 ten. One hundred equals 10 tens; 10 + 1 = 11. • Subtract 3 tens from 11 tens, leaving 8 tens. • When 1 hundred was taken from 8 hundreds, 7 hundreds were left. • Subtract 5 hundreds from 7 hundreds, leaving 2 hundreds. The remainder is 289 sq. ft.

CHECKING SUBTRACTION

To check subtraction, add the difference, or remainder, to the subtrahend. If the subtraction is correct, the sum will be the same as the minuend.

Illustration. A contractor paid to have 933 cubic yards (cu. yd.) of dirt excavated. He found later that only 419 cu. yd. had been moved. How much dirt was not excavated?

Solution.

			Check:		
	933	minuend		514	remainder
−	419	subtrahend		419	subtrahend
	514	remainder		933	sum = minuend

Time-saving check:

add ⎰ 933 minuend ⎱ subtract
419 subtrahend
514 remainder
933 minuend

Explanation. • Since the sum of the subtrahend (419) and the remainder (514) equals 933, the subtraction is correct. To save time in checking, it is not necessary to rewrite the remainder (514) and the subtrahend (419).

COMBINING ADDITION AND SUBTRACTION

Many times, a carpenter must solve problems that require both addition and subtraction.

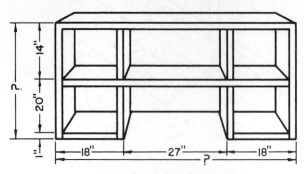

Fig. 3-3 Cabinet

Illustration. A carpenter is working on a kitchen cabinet, figure 3-3. The height of the cabinet is marked in sections of 1″, 20″, and 14″. The length of the cabinet is marked 18″, 27″, and 18″. What is the difference between the height and length of the cabinet?

Solution.   Add   Subtract

	18″		1″		63″
	27″		20″		35″
	18″		14″		28″ difference in height
	63″ sum of length		35″ sum of height		and length

Explanation. • Add all height measurements to find the total height: 1″ + 20″ + 14″ = 35″. • Add 18″, 27″, and 18″ to obtain the total length, 63″. • After the total height and length are determined, subtract 63″ − 35″ = 28″.

APPLICATION

1. How many feet will remain on a piece of pipe that is 18′ long if a 4-foot piece is cut off?

2. Find the difference between each set of numbers.

9	4	3	4	6	7	5	5	7
- 3	- 4	- 3	- 2	- 3	- 2	- 3	- 2	- 5

7	8	9	9	9	9	6	6	6
- 3	- 3	- 7	- 5	- 4	- 6	- 1	- 4	- 5

3. Find the difference between each set of numbers.

3 − 3 =	8 − 6 =	6 − 2 =	7 − 3 =	9 − 3 =
6 − 6 =	6 − 5 =	7 − 1 =	8 − 2 =	5 − 2 =
5 − 1 =	9 − 5 =	5 − 4 =	8 − 0 =	8 − 3 =

4. Subtract each pair of numbers.

12	17	12	13	11	18
- 9	- 6	- 7	- 4	- 8	- 6

15	14	11	16	13	11
- 9	- 2	- 4	- 5	- 5	- 6

5. Subtract each pair of numbers.

87	28	295	954	2672	3204
- 26	- 19	- 164	- 665	- 1938	- 2019

685	401	493	509	1001	4200
- 93	-385	- 387	- 327	- 890	- 3299

6. A standard outside door measures 3′ wide and 7′ high. The standard height of a wall is 8′. How much higher is the wall than the door?

7. A truck is loaded with 1658 bricks. After 489 are unloaded, how many bricks will be left on the truck?

8. The measurements of the room shown in figure 3-4 are as follows: 16′ on the north side, 16′ on the south side, 12′ on the east side and 12′ on the west side. What is the difference between the north-south measurements and the combined east-west measurements?

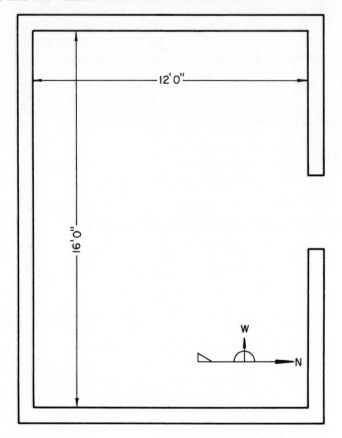

9. The height of a wall is 108″ and the height of a door is 68″. How many inches higher is the wall than the door?

10. Subtract and check the following problems.

365	275	337	395	583	691
- 334	- 245	- 306	- 274	- 274	- 219

11. Arrange and subtract each of the following sets of numbers.

84 and 56	3,650 and 5,850
35 and 49	4,859 and 5,948
58 and 99	3,945 and 4,539
35 and 67	243,870 and 879,660
85 and 32	326,050 and 85,914
87 and 49	175,925 and 35,467

Unit 4
Multiplication of Whole Numbers

OBJECTIVES

After studying this unit, the student should be able to

- multiply numbers with two or more digits.
- solve problems that require both multiplication and subtraction.
- use multiplication in carpentry.
- check problems for accuracy.

Multiplication is a simplified method of adding a quantity a given number of times. For example, when 2 is multiplied by 4, the product is the same as in the process $2 + 2 + 2 + 2$. In the process of multiplication, the number multiplied is the *multiplicand*. The number by which the multiplicand is multiplied is the *multiplier*. The result is called the *product*. Multiplication may be shown either horizontally or vertically.

Illustration. The crosscut handsaw is designed to cut across the grain of wood. It is identified by its shape and the number of teeth per inch. A crosscut saw blade is 26″ long and has 8 teeth per inch, figure 4-1. Find the number of teeth on the saw blade.

Solution.
$$\begin{array}{r} 26 \\ \times\ 8 \\ \hline 208 \end{array} \begin{array}{l} \text{multiplicand} \\ \text{multiplier} \\ \text{product} \end{array}$$

Explanation. • The multiplicand is 26, the multiplier is 8 and the product is 208 teeth.

MULTIPLYING SMALL WHOLE NUMBERS

When multiplication of more than two small whole numbers is required, find the product of the first two numbers and then the final product.

Illustration. Find the number of concrete blocks in figure 4-2.

Solution. $5 \times 6 \times 7 = 210$

Explanation. • Multiply 5×6 to obtain 30, and multiply 30×7 to obtain the final product (210).

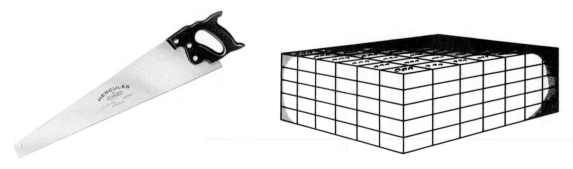

Fig. 4-1 Crosscut Handsaw

Fig. 4-2 Concrete Blocks

SHORTCUTS IN MULTIPLICATION

- The product of zero times another number is zero: $0 \times 0 = 0$, $0 \times 5 = 0$, $0 \times 9 = 0$.
- To multiply a number by 10, annex 0: $21 \times 10 = 210$.
- To multiply a number by 100, annex 00: $48 \times 100 = 4,800$.
- To multiply a number by 1000, annex 000: $52 \times 1000 = 52,000$.
- To multiply a number by another number which is greater than 1 and ends in one or more zeros, annex the zeros and multiply the numbers which are left: $27 \times 200 = 5400$.

How To Carry Numbers

The carrying process used in addition is discussed in Unit 2. This carrying process involves placing addends in vertical order with units under units, tens under tens, and so forth. A similar process is used in multiplication.

Illustration. *Framing studs* are vertical members of outside and partition walls which support the weight of the upper sections and provide a frame for fastening sheathing. The standard length of a stud, figure 4-3, is 96″. How many inches of material are required to make 6 studs?

Solution.
$$\begin{array}{r} 96'' \\ \times\ 6 \\ \hline 576'' \end{array}$$

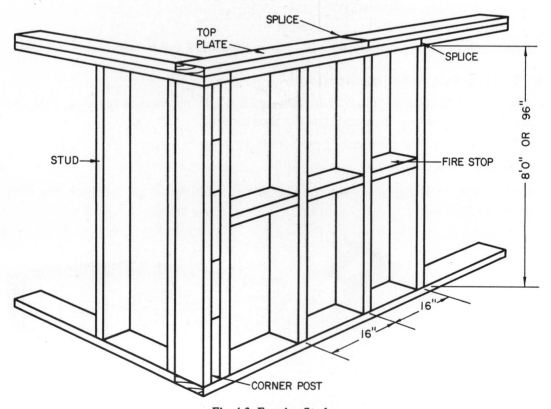

Fig. 4-3 Framing Studs

Explanation. • Multiply 6 x 6 for a product of 36. Write 6 under the line in the units column and carry the 3. • Multiply 6 x 9 for a product of 54 and add the 3 which was carried: 54 + 3 = 57. Write 57 to the left of 6. • It takes 576″ of material to make 6 studs.

One number may be multiplied by a multiplier which has a zero between two other figures.

Illustration. *Footings* are concrete bases upon which foundation walls and piers are built. Footings provide support and stability for structures. The concrete footings in figure 4-4 share equally in supporting the weight of a building. If each of 105 footings support 284 lb., what is the weight of the building?

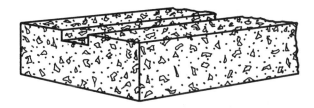

Foundation Wall Footing   **Pier Footing**

Fig. 4-4

Solution.   284 lb.
                 105 footings
             1420
           284
           29820 lb.

Explanation. • Notice that the 4 is placed under the hundreds column when multiplying by 1. A zero indicating the product of 0 x 284 may be placed in the tens column.

An acceptable way of checking for accuracy is to exchange places of the multiplicand and the multiplier.

Illustration. Exchange the multiplier and multiplicand.

Solution.   67 multiplicand   *Check:*   16 multiplicand
            16 multiplier   67 multiplier
         402   112
          67   96
         1072 product   1072 product

Explanation. • The same product is obtained in both problems. By rearranging numbers, errors are more likely to be found.

APPLICATION

1. Multiply the following sets of numbers.

a. 45 x 9	f. 34 x 5	k. 53 x 9	o. 62 x 2	s. 28 x 67	w. 38 x 44
b. 83 x 4	g. 87 x 9	l. 30 x 7	p. 79 x 5	t. 3242 x 76	x. 1036 x 230
c. 58 x 6	h. 41 x 8	m. 95 x 9	q. 86 x 4	u. 342 x 300	y. 284 x 403
d. 76 x 3	i. 91 x 4	n. 82 x 9	r. 358 x 26	v. 752 x 305	z. 487 x 89
e. 92 x 7	j. 70 x 7				

2. A carpenter works 40 hours per week. How many hours does he work in 42 weeks?

3. One row of shelves in a kitchen cabinet measure 58 sq. ft. How many square feet do 19 rows measure?

4. A carpenter has 38 pieces of wood for studs. If each piece is 96″ long, what is the total amount used for studding?

5. A builder buys 23 second-hand saws at $3.00 each. He overhauls the saws and sells them for $8.00 each. What is his total profit?

6. Using the figures shown, design footings for a framed construction and a masonry construction. Indicate thickness and width. *Note:* Hard rock soil and sand and gravel soil support 2 tons per square foot. Loam soil supports 1 ton per square foot. The thickness of a footing should equal the thickness of the foundation wall.

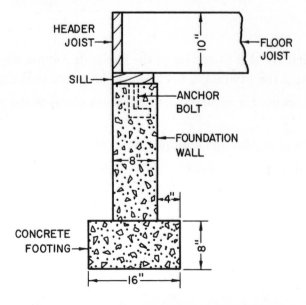

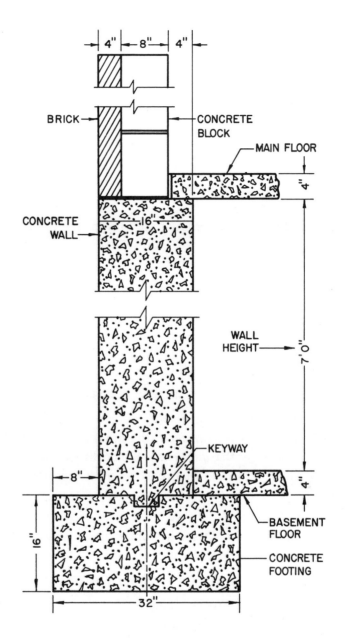

7. A carpenter wants to purchase equipment totaling $1150. He can pay $350 down in cash and pay off the balance in 24 monthly installments of $40 each. How much could he save by paying the entire sum at the time of purchase?

Unit 5
Division of Whole Numbers

OBJECTIVES

After studying this unit the student should be able to

- divide whole numbers using the long and short methods.
- solve problems in carpentry requiring the use of division.
- check division problems for correctness.

When one of two factors and the product are known, the procedure used to find the missing factor is division, such as, $5 \times \_ = 20; 20 \div 5 = 4$. *Factors* are numbers which may be multiplied together to produce a given number.

Just as multiplication is a short method of addition, division may also be thought of as a short method of subtraction. Whether 5 is subtracted from 20 four times, or 20 is divided by 5, the result is the same:

$$20 - 5 - 5 - 5 - 5 = 0$$

$$20 \div 5 = 4$$

$$4 \times 5 = 20, \text{ with 0 remaining}$$

The purpose of division is to find how many times one number contains another number. The number to be divided is the *dividend;* the number that is used to divide by is the *divisor.* The result of division is the *quotient.* Some division quotients are exact; some leave *remainders,* that part of the dividend left over when the quotient is not exact.

SHORT DIVISION

Short division is used when the divisor has one digit or when the only other digits are zeros. The dividend may contain several digits.

Illustration. How many 4' pieces can be cut from 476' of material?

Solution.

$$\text{divisor } 4' \overline{)476'} \text{ dividend} \qquad \frac{119}{} \text{ quotient}$$

Explanation. A number must be determined which, when multiplied by the divisor, gives a product equal to the first number in the dividend. If such a number cannot be found, one which is as near equal as possible but not larger than the number must be determined. • In the example, 4 goes into 4 exactly one time. The number 1 is written in the quotient above the line directly over the 4. • The next number (7) is then divided by 4. In this instance, 4 goes into 7 one time with 3 left over. The 1 is written above the 7. • Mentally place the remainder (3) in front of the 6, making the figure 36. • The 4 is then divided into 36 and the result (9) is written in the quotient directly above the 6. • The answer is 119 pieces, each 4 ft. long. In this problem, 4 is an exact divisor of 476.

LONG DIVISION

It is convenient to use the long division method when the divisor contains two or more digits.

Illustration. There are 32 sq. ft. in one piece of 4' x 8' plywood, figure 5-1. How many pieces are there in 834 sq. ft.?

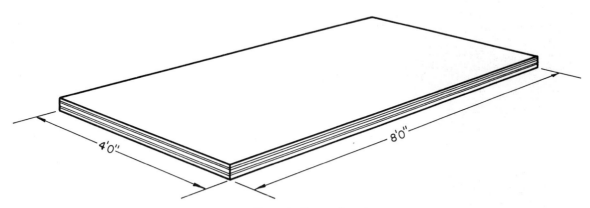

Fig. 5-1 Plywood

Solution.

```
                         26 pieces quotient
         divisor  32 sq. ft ) 834 sq. ft.  dividend
                         64
                         ───
                         194
                         192
                         ───
                           2 remainder
```

Explanation. In the problem, 32 is not contained in 8, but it is contained 2 times in 83. • Write 2 in the quotient above 3 in the dividend and multiply 32 by 2. • Write the product (64) under 83. Subtracting 64 from 83 gives a remainder of 19. • Bring down the 4 and write it beside 19, making the number 194. • Finally, 32 is contained in 194 about 6 times. • Write 6 in the quotient above 4 in the dividend and multiply 32 by 4. • Write the product (192) under 194. Subtracting 192 from 194 leaves a remainder of 2. The answer is 26 pieces with 2 as the remainder. In this case, 32 is not an exact divisor of 834.

MEASURING FOR EQUAL DIVISION

Dressed lumber measures less than rough, or unfinished, lumber. The measurement of lumber is always expressed in its rough size. *Dressed* lumber has been planed on its edges and surfaces.

Figure 5-2 shows how a piece of 1" x 10" wood is divided into 3 equal parts. The rule is slanted across the surface of the board until it can be divided into the desired equal number of spaces from the divisions on the rule.

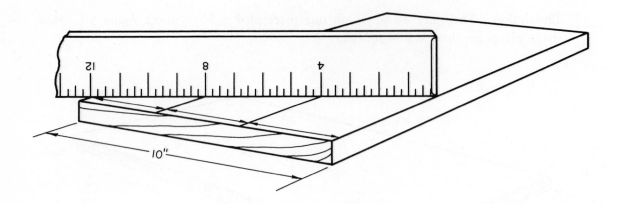

Fig. 5-2 Measuring for Equal Division

If the board is to be divided into an odd number of pieces, use any number on the rule divisible by the desired number of pieces and mark points accordingly. In the illustration, the number 12 is used because it is divisible by 3, an odd number. Since $12 \div 3 = 4$, points 4 and 8 are marked on the board. When the board is divided into an even number of pieces, use even numbers on the rule.

CHECKING DIVISION

To check results in division, multiply the divisor by the quotient and add the remainder. If the result is the same number as the dividend (original number), the division is correct.

Illustration. Studs are to be placed 16″ on center (O.C.) for a wall 41′ long. How many studs are needed?

Note: On center is a term used in carpentry to indicate the spacing of framing members from the center of one member to the center of the next.

Solution. 41 x 12″ = 492″ wall 　　　　30 spaces
　　　　　Divide 492 by 16.　　　　　　+1 remainder (requires 1 space)
　　　　　492 ÷ 16 = 30　　　　　　　　31 spaces
　　　　　　　　　　　　　　　　　　　+1 (as a starter)
　　　　　　　　　　　　　　　　　　　32 studs
　　　　Check.　16 x 30 + 12 = 492

Explanation. • Multiplication is used to change 41′ to inches: 12 x 41 = 492″. • Then, 492 is divided by 16. • The quotient indicates that there are 30 spaces which are 16″ O.C. and one 12″ space. • The 12″ space requires a stud and one stud is needed as a starter; therefore, a total of 32 studs is needed.

APPLICATION

1. Identify the missing number in each of the following. Indicate the dividend, divisor, quotient, and remainder in each.

   a. 8 x _ = 32   b. _ x 12 = 72   c. 9 x _ = 105

2. Divide each of the following by short division.

   a. 30 ÷ 5   d. 69 ÷ 4   g. 31 ÷ 5
   b. 87 ÷ 5   e. 83 ÷ 3   h. 61 ÷ 7
   c. 15 ÷ 5   f. 78 ÷ 6

3. A *lintel* is a piece of material which is placed over each window opening of a structure to support the walls above the opening. Lintels for 28 equal windows are cut from 74 feet of material. How long is each piece?

4. A crosscut saw is equipped with a blade 26″ long with a total of 208 teeth. How many teeth are there per inch?

5. Divide each of the following using long division.

   a. 156 by 90   e. 155 by 31   i. 2385 by 22
   b. 366 by 153   f. 342 by 18   j. 550 by 403
   c. 455 by 69   g. 169 by 13   k. 4218 by 207
   d. 365 by 89   h. 924 by 28   l. 36315 by 423

6. Arrange and divide the following numbers. Check for accuracy.

   a. 587 ÷ 27   d. 300 ÷ 3   g. 9,275 ÷ 128
   b. 1,221 ÷ 3   e. 1,255 ÷ 5   h. 2,439,122 ÷ 1,233
   c. 1,223 ÷ 13   f. 4,226 ÷ 6   i. 4,918,842 ÷ 4,918

7. Solve each of the following problems. They require a combination of processes.

   a. (5 + 5 + 10 – 4) ÷ 2   c. [(12 x 12 x 12) – 144] ÷ 4
   b. (15 + 18 + 2) x 6 ÷ 5   d. [(108 x 145) + 55] ÷ 100

8. A concrete form for a pier requires several pieces of 2″ x 8″ lumber 2 feet long. How many pieces can be cut from 2 sections of lumber measuring 2″ x 8″ x 10′?

9. The concrete form in the figure on page 26 is constructed of 3/4-inch thick, 4′ x 4′ plywood, sawed in four equal widths. How many inches wide is each of the four pieces?

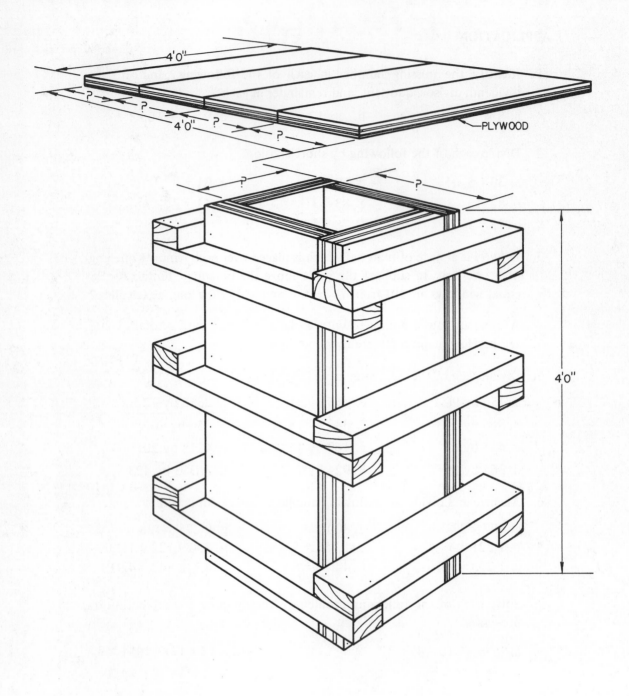

10. How many rolls are there in 2500′ of Number 12 electrical wire if each roll consists of 250′ of wire?

11. A structure has 50 electrical outlets which must be wired. The city code allows a maximum of 5 outlets to a circuit and requires the switch box to be large enough to carry space for 2 extra circuit breakers. How many circuit breakers should the switch box shown in the figure below hold?

12. Divide a piece of 1″ x 10″ lumber into 4 equal pieces by slanting a rule across the stock.

13. Divide a piece of 1″ x 8″ lumber into 3 equal widths by slanting the rule across the stock.

14. A plumbing job calls for 800′ of 1″ pipe. Allowing 20 feet for each joint, how many joints are there?

CAREER PROFILE: THE FOUNDATION OR FORM BUILDER

Carpenters who construct foundations are the first workers to begin construction on a building. They lay out foundations for new constructions or additions to existing structures.

Job Description

Specifically, form builders lay out building lines, erect batter boards, construct wall footing forms, pier footing forms, wall forms, and concrete step forms. Much of their job is spent reading blueprints and specifications. Outdoor work offers varied conditions in this profession.

Qualifications

As in most carpentry jobs, preference is given to high school graduates or vocational-technical school graduates. Generally, they complete a two- or three-year apprenticeship program. Besides specific study for the job, courses in related mathematics and blueprint reading are essential. Good physical condition is necessary.

Section 2
Fractions

Unit 6
Addition of Fractions

OBJECTIVES

After studying this unit, the student should be able to

- add fractions with common denominators.
- add fractions with different denominators.
- change improper fractions to whole or mixed numbers.

A fraction is a part of a whole number. One is the smallest whole number. One-half is the largest equal fraction; one-third is the next largest with one-fourth following.

When expressed in words, fractions are compound words and are hyphenated.

One foot is divided into 12 fractional parts called inches; each inch is divided into many fractional parts, figure 6-1.

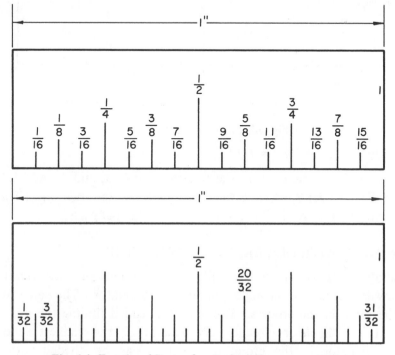

Fig. 6-1 Fractional Parts of an Inch as Shown on a Rule

The rule or ruler is probably the most frequently used measuring tool in carpentry. The ability to read the fractional parts of a rule is necessary for anyone working in this field.

Examine the divisions of an inch on a rule and notice the length of each mark. The longest line indicates 1/2"; the next, 1/4" and 3/4"; the next, 1/8", 3/8", 5/8" and 7/8".

The number above the horizontal line in fractions is the *numerator;* the number below the horizontal line is the *denominator.* The denominator shows how many parts into which a whole object has been divided. The numerator indicates the number of equal parts of the denominator which are taken.

$$\text{numerator} \longrightarrow \frac{1}{2} \longleftarrow \text{denominator}$$

ADDING FRACTIONS WITH COMMON DENOMINATORS

In the addition of fractions, the sign (+) or the word *plus* is used, as in the addition of whole numbers. The process of combining two or more fractions with identical denominators is known as addition of fractions. Only fractions with a same, or *common,* denominator can be added: 1/3 + 1/3 = 2/3.

When adding fractions having a common denominator, add the numerators and write the sum over the denominator.

Illustration. Using figure 6-2, find the sum of 1/8", 3/8" and 3/8".

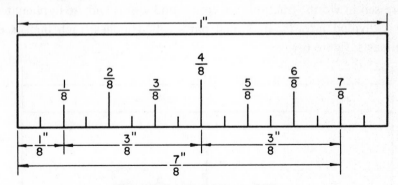

Fig. 6-2 Fractional Parts of 1"

Solution. 1/8" + 3/8" + 3/8" = 7/8"

Explanation. • The numerators in the fractions are 1, 3, and 3. Add 1 + 3 + 3 which equals 7. • Place 7 over 8, which is the denominator of the fraction. • The sum is 7/8" which indicates that there are 7 parts taken out of 8 equal parts.

ADDING FRACTIONS WITH DIFFERENT DENOMINATORS

When adding fractions which have different denominators, the fractions must be changed to equivalent fractions having a common denominator. The value of the fraction must remain the same in the process. The process is called finding the least common denominator (L.C.D.). The *least common denominator* is the smallest denominator into which each denominator in the group of numbers can be divided without leaving a remainder.

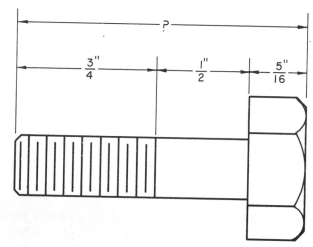

Fig. 6-3 Bolt

Illustration. What is the total length of the bolt in figure 6-3?

Solution.

$$\frac{1''}{2} = \frac{8''}{16}$$

$$\frac{3''}{4} = \frac{12''}{16}$$

$$\frac{5''}{16} = \frac{5''}{16}$$

$$\frac{25''}{16} = 1\ \frac{9''}{16}$$

Explanation. The smallest number into which 2, 4, and 16 can be divided without leaving a remainder is 16. Therefore, 16 is the least common denominator. Change each fraction to its equivalent in 16ths. The sum of 8, 12, and 5 (25) is placed over the denominator (16) to form the answer, 25/16″ or 1 9/16″.

CHANGING AN IMPROPER FRACTION TO A WHOLE OR MIXED NUMBER

Fractions may be proper or improper. When the fraction has a denominator larger than the numerator, such as 1/2, 1/4, 2/3, or 3/8, it is a *proper* or *common* fraction. If the numerator of a fraction is larger than the denominator, such as 25/16, it is an *improper* fraction. A *mixed number* is a number which is the sum of a whole number and a proper fraction.

At times, improper fractions must be changed to whole or mixed numbers. To change an improper fraction to a whole number, divide the numerator by the denominator: 25 ÷ 16 = 1 9/16. In this case, the fraction 25/16 is greater than one whole. The remainder (9) is the numerator of the fractional part of the mixed number, 1 9/16. The denominator (16) is the same as in the improper fraction, 25/16.

REDUCING FRACTIONS TO LOWEST TERMS

A fraction is in lowest terms when both the numerator and denominator cannot be divided by the same number. To reduce a fraction to its lowest terms, divide both numerator and denominator by the same number an equal number of times. The value of the fraction does not change.

Illustration. The length of a layout is expressed as 20/32. Reduce 20/32 to its lowest terms using the cancellation method.

Solution.

$$\frac{\overset{\overset{5}{\cancel{10}}}{\cancel{20}}}{\underset{\underset{8}{\cancel{16}}}{\cancel{32}}} = \frac{5}{8}$$

Explanation. • Two will divide into both 20 and 32 an exact number of times (20 ÷ 2 = 10: 32 ÷ 2 = 16). • Mark a line through 20 and 32 and write the quotients. • Again, 2 will divide into both 10 and 16 an exact number of times (10 ÷ 2 = 5; 16 ÷ 2 = 8). No number greater than 1 will divide into both 5 and 8 an exact number of times, so the fraction 5/8 is reduced to its lowest possible terms. • The value of 5/8 is exactly the same as the value of 20/32.

When solving problems involving fractions, the student must sometimes change denominators to match a given, or designated, denominator. If both the numerator and denominator of a fraction are multiplied or divided by the same number, the value of the fraction remains unchanged.

To change a fraction to higher terms with a given denominator, divide the required higher denominator by the denominator of the fraction and multiply both the numerator and denominator of the fraction by the quotient.

Illustration. Change 3/8 to twenty-fourths.

Solution. 24 ÷ 8 = 3
3/8 x 3/3 = 9/24

Explanation. • Divide the required denominator (24) by the original denominator (8) to obtain the quotient (3). • Multiply the numerator (3) by the quotient (3) to obtain the numerator for the new fraction (9). • Multiply the denominator (8) by the quotient (3) to obtain the denominator for the new fraction (24). • The new fraction is 9/24.

To reduce a fraction to lower terms with a given denominator, divide the higher denominator by the required lower denominator of the fraction and divide both the numerator and denominator by the quotient.

Illustration. 21/35 = ?/5

Solution. 35 ÷ 5 = 7
21/35 ÷ 7/7 = 3/5

Explanation. • Divide the original denominator (35) by the required denominator (5) to obtain the quotient (7). • Then divide the numerator (21) and the denominator (35) by 7. • The new fraction is 3/5. • Remember that 21/35 and 3/5 are equal in value.

APPLICATION

1. Reduce the following fractions to lower terms.

a. 2/8	g. 3/15	m. 10/40	s. 35/45
b. 4/8	h. 16/32	n. 12/52	t. 12/72
c. 3/9	i. 5/10	o. 12/42	u. 18/76
d. 4/6	j. 6/10	p. 12/48	v. 50/120
e. 10/12	k. 7/21	q. 8/20	w. 75/125
f. 14/16	l. 4/32	r. 30/44	

2. Reduce the following fractions to lowest terms.

a. 6/8	e. 2/8	i. 40/120	m. 12/36
b. 6/12	f. 8/10	j. 16/24	n. 25/50
c. 4/12	g. 6/9	k. 5/125	
d. 14/16	h. 3/21	l. 12/24	

3. On a sheet metal layout, 5 holes are drilled as in the figure below. What is the total length of the piece?

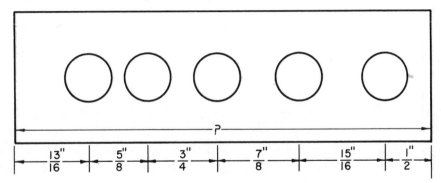

4. A *dovetail* joint, constructed by interlocking two boards, is illustrated in the figure below. What is the width of the board in the figure?

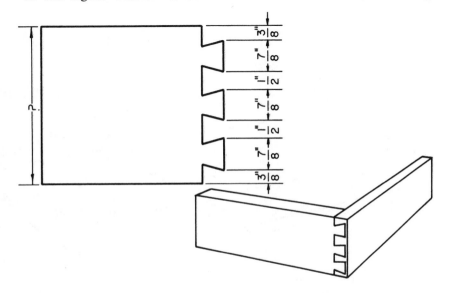

5. Add the following fractions and reduce answers to lowest terms.

   a. 1/2 + 1/2 =   d. 1/6 + 1/6 =
   b. 1/4 + 1/4 =   e. 1/2 + 1/4 =
   c. 1/8 + 1/8 =   f. 1/4 + 1/8 =

6. Add the following fractions, changing the sum to a whole or mixed number if necessary.

   a. 1/6 + 1/2 + 7/8 + 3/32 =   e. 5/9 + 5/12 + 11/36 + 5/48 =
   b. 1/8 + 3/16 + 7/16 + 5/12 =   f. 1/2 + 5/8 + 3/5 + 3/4 =
   c. 3/8 + 3/4 + 5/32 + 14/16 =   g. 5/11 + 5/13 + 6/11 + 6/13 =
   d. 3/21 + 5/14 + 5/7 + 5/28 =

7. The *finger* joint, a commonly used joint in construction, is illustrated below. What is the total width of the joint in the figure?

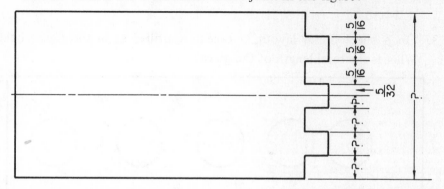

8. The *dado* joint, as shown in the figure below, is usually formed by the intersection of two boards placed at a right angle to each other. What is the thickness of the groove used to form the joint in the figure?

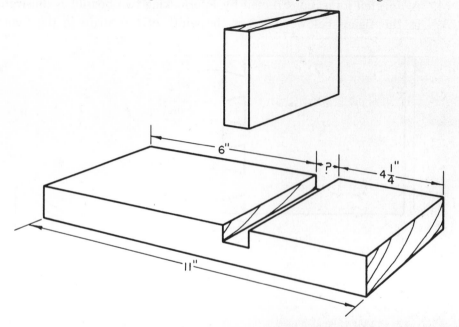

9. Pipes and copper tubing used by plumbers are available in standard sizes such as 3/8", 1/4", 1/2", 5/8", and 3/4". The sizes are according to the measurement of their inside diameters. If all these sizes were added to make one diameter, what would the size be?

10. A carpenter purchases several sizes of nails: 1/4 lb. of brads, 1/2 lb. of number 4 nails, 19/32 lb. of number 6d finishing nails, and 13/16 lb. of number 8d common nails. How many pounds of nails did he purchase?

11. Change the following fractions to higher terms, using the denominator given.

    a. 5/6 = ?/30   f. 5/8 = ?/16   k. 3/9 = ?/27   p. 5/8 = ?/32
    b. 3/4 = ?/12   g. 9/24 = ?/72   l. 1/4 = ?/64   q. 1/8 = ?/16
    c. 3/6 = ?/12   h. 3/9 = ?/12   m. 1/2 = ?/12   r. 15/32 = ?/64
    d. 15/16 = ?/32   i. 9/24 = ?/48   n. 3/6 = ?/12   s. 1/2 = ?/30
    e. 1/2 = ?/120   j. 11/16 = ?/64   o. 19/32 = ?/64   t. 7/10 = ?/100

Unit 7
Subtraction of Fractions

OBJECTIVES

After studying this unit, the student should be able to

- subtract fractions with common denominators.
- subtract fractions with different denominators.
- solve problems that involve both addition and subtraction of fractions.

Before one fraction can be subtracted from another, the fractions must have common denominators. Sometimes a fraction must be reduced to lower or lowest terms.

SUBTRACTION OF FRACTIONS WITH COMMON DENOMINATORS

To subtract fractions with common denominators, subtract the smaller numerator from the larger numerator. Write the difference above the division line and the common denominator below the line.

Illustration. A *parting strip*, figure 7-1, is a piece of wood which separates two sashes of a window. The parting strip of a window is 7/8″ wide, but is later cut down by 3/8″. What is the new width?

Solution. $7/8'' - 3/8'' = 4/8'' = 1/2''$

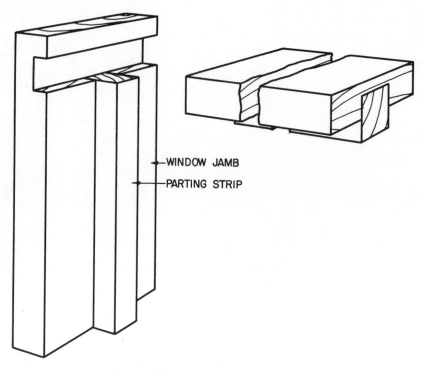

WINDOW JAMB

PARTING STRIP

Fig. 7-1 Window Jamb With Parting Strip

Explanation. • Write the larger fraction (7/8″) first and place the minus sign (−). • Write the other fraction (3/8″). • Subtract the smaller numerator (3) from the larger numerator (7). • Write the difference (4) above the line over the common denominator (8). • The new width is 4/8″.

   • Reduce 4/8″ to its lowest terms by dividing both the numerator and denominator by 4: $4/8 \div 4/4 = 1/2$. The 1 and 2 are said to be prime to each other, so the fraction 1/2 is in its lowest terms. Two or more numbers are *prime* to each other when they cannot both be divided an exact number of times by the same number.

   The fraction 4/8 can also be reduced by the cancellation method:

$$\frac{\overset{\overset{1}{\cancel{2}}}{\cancel{4}}}{\underset{\underset{2}{\cancel{4}}}{\cancel{8}}} = \frac{1}{2}$$

SUBTRACTION OF FRACTIONS WITH DIFFERENT DENOMINATORS

Different denominators in fractions must be converted to a common denominator before one can be subtracted from the other. After obtaining a common denominator, subtract the smaller numerator from the larger. Write the difference in the numerators above the division line and the common denominator below the line.

Illustration. A board 11/16″ thick is planed to a thickness of 1/2″, figure 7-2. How much is removed? *Note:* At this stage 1/2″ cannot be subtracted from 11/16″ because the denominators are not the same. The fraction 1/2 is converted to a fraction with 16 as the denominator.

Solution. Step 1. $16 \div 2 = 8$
   Step 2. $1/2 = 1/2 \times 8/8 = 8/16$
   Step 3. $11/16 - 8/16 = 3/16$

Explanation. • Determine the common denominators. • Divide 2 into 16 to obtain the quotient (8). • Multiply the numerator (1) by the quotient (8) to obtain the numerator (8). • Multiply the denominator (2) by the quotient (8) to obtain the denominator (16). • The fraction 1/2 is converted to 8/16. • Subtract 8/16 from 11/16.

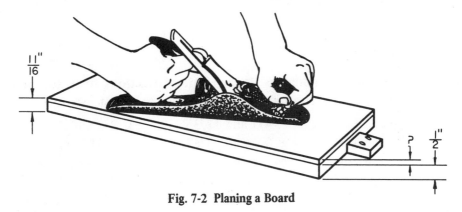

Fig. 7-2 Planing a Board

SUBTRACTION OF FRACTIONS FROM WHOLE NUMBERS

To subtract a proper fraction from a whole number, change one unit of the whole number to a fraction which is equal to the unit and which contains the common denominator.

Illustration. Subtract 7/8 from 8.

Solution.   8 – 7/8 = __

7 8/8 – 7/8 = 7 1/8

Explanation. • Change one unit of the whole number (8) to a fraction which is equal to the unit and which contains the common denominator (8/8). • The whole number (8) is equal to 7 8/8. • The fractions are now subtracted as usual.

COMBINING ADDITION AND SUBTRACTION OF FRACTIONS

Many times, problems in carpentry involve both addition and subtraction of fractions.

Illustration. From the sum 1/4 and 5/8, subtract the difference between 1/2 and 3/16.

Solution.   Step 1.   Add 1/4 and 5/8.   1/4 = 2/8
                                                    + 5/8 = 5/8
                                                       7/8 sum

Step 2.   Subtract 3/16 from 1/2.   1/2 = 8/16
                                      – 3/16 = 3/16
                                         5/16 difference

Step 3.   Subtract 5/16 from 7/8.   7/8 = 14/16
                                      – 5/16 = 5/16
                                            9/16

Explanation. • Find the sum of the two fractions (1/4 and 5/8): 2/8 + 5/8 = 7/8. • Secondly, find the difference between 1/2 and 3/16: 8/16 – 3/16 = 5/16. • Change 7/8 to 14/16. • Subtract the difference (5/16) from 14/16. • The result is 9/16.

CHECKING SUBTRACTION OF FRACTIONS

To check a fraction, add the subtrahend to the difference. The result is correct if this sum is equal to the minuend. For example, in the problem 14/16 – 5/16 = 9/16, add the subtrahend, 5/16, and the difference, 9/16. The result is 14/16, which is equal to the minuend.

APPLICATION

1. Change the denominators in each pair of fractions to least common denominators and subtract. Reduce answers to lowest terms.

a. 5/6 and 1/8	f. 4/7 and 12/35	k. 3/5 and 4/7
b. 1/2 and 5/64	g. 9/16 and 3/32	l. 11/16 and 3/8
c. 4/5 and 3/6	h. 7/8 and 9/16	m. 7/16 and 1/4
d. 3/16 and 3/64	i. 61/64 and 21/32	n. 7/8 and 3/5
e. 9/16 and 3/8	j. 7/8 and 1/16	o. 11/13 and 5/11

2. In constructing the dado joint in the figure below, one board is grooved 9/16″. The board to be fitted in the groove is 28/32″ thick. To join these boards together, how much larger should the groove be?

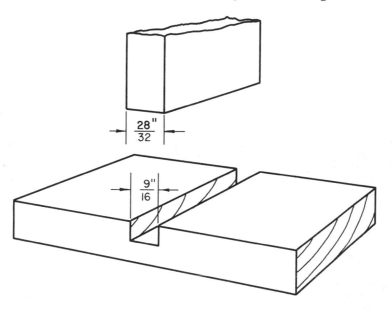

3. Wooden dowels are used to add strength to various kinds of butt joints. The dowels in the figure below measure 1/2″ and 7/8″. What is the difference in their sizes?

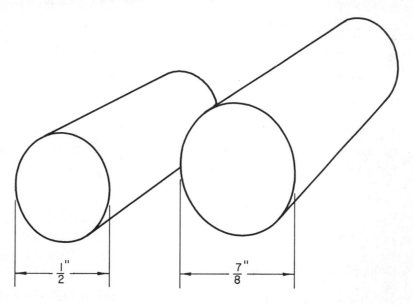

4. A rabbet is cut 9/16" deep in a piece of material 29/32" thick in the figure. What is the thickness of the remaining portion?

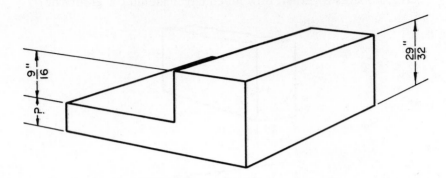

5. Arrange the following pairs of fractions and subtract. If necessary, first find the least common denominator.

a. 1/2 and 3/5	h. 13/15 and 1/4	o. 19/24 and 7/12
b. 7/8 and 1/3	i. 9/10 and 3/8	p. 7/8 and 3/4
c. 1/6 and 1/9	j. 9/16 and 17/20	q. 3/5 and 5/8
d. 3/5 and 9/10	k. 2/3 and 2/5	r. 13/20 and 3/5
e. 4/5 and 3/4	l. 13/18 and 7/12	s. 5/6 and 3/7
f. 7/10 and 3/8	m. 1/6 and 2/3	t. 13/15 and 4/5
g. 3/5 and 1/4	n. 1/4 and 3/5	u. 5/8 and 4/5

6. Use the cancellation method to reduce the following fractions to their lowest terms.

a. 10/120	h. 38/76	o. 16/32
b. 55/75	i. 3/15	p. 69/69
c. 10/50	j. 6/8	q. 2/4
d. 42/60	k. 20/32	r. 20/40
e. 25/150	l. 32/64	s. 16/48
f. 3/45	m. 12/12	t. 45/55
g. 40/64	n. 16/16	

7. The nominal size of a piece of hardwood flooring is 1/2" thick and the actual size is 15/32" thick. What is the difference in thickness? *Note: lumber is referred to in both nominal and actual sizes. Nominal is the commercial size given by the trade; actual is the mill size. They are not always the same size.*

8. The nominal size of a piece of hard maple flooring is 3/8" thick and the actual size is 11/32" thick. What is the difference in thickness?

9. For general work, the carpenter frequently uses auger bits ranging from 1/4" to 1" in size. How much larger is the hole that a 1" bit bores than a 1/4" bit?

10. Wood chisels, shown in the figure below, are cutting tools with many uses. Chisels come in a variety of sizes and styles. What is the difference in width between a 3/4" and 3/8" chisel?

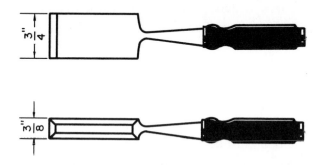

Unit 8
Multiplication of Fractions

OBJECTIVES

After studying this unit, the student should be able to

- multiply fractions by other fractions.
- multiply whole or mixed numbers by fractions.

Multiplication of fractions occurs frequently in carpentry work. Remember that multiplication is simply a short method of addition.

MULTIPLYING A FRACTION BY ANOTHER FRACTION

The multiplication of fractions does not require finding the least common denominator. When one fraction is multiplied by another, the numerators are multiplied together to form one numerator and the denominators are multiplied together to form one denominator. Sometimes the new fraction must be reduced to its lowest terms.

Illustration. Multiply 2/5 x 5/12 and reduce to lowest terms.

Solution.   2/5 x 5/12 = 10/60

$$\frac{\overset{\overset{1}{\cancel{5}}}{\cancel{10}}}{\underset{\underset{6}{\cancel{30}}}{\cancel{60}}} = \frac{1}{6}$$

Explanation. • Multiply the numerators (2 x 5) and write the product (10) as the new numerator. • Multiply the denominators (5 x 12) and write the product (60) as the new denominator. • The product is 10/60. Reduce 10/60 to its lowest terms. • The new product is 1/6.

MULTIPLYING A WHOLE NUMBER BY A FRACTION

To multiply a whole number by a fraction, multiply the whole number by the numerator of the fraction. Then, divide that product by the denominator of the fraction. If the product is an improper fraction, reduce it to a whole or mixed number.

Illustration. How many feet are there in 9 pieces of pipe, each 2/3′ long?

Solution.   9 x 2/3′ = 18/3′ = 6′

                        or

       9/1 x 2/3′ = 18/3′ = 6′

Explanation.   • Assume that all whole numbers have 1 as a denominator. • Multiply the numerators (9 x 2 = 18) and the denominators (1 x 3 = 3). • Then, 18/3 is changed to a whole number (6).

42

PRACTICAL APPLICATION

Determining The Rise of a Roof

The *span* of a roof is the distance from the seat cut of one rafter to the seat cut of the opposite rafter. In equally-pitched roofs, the *run* is equal to half the span or, generally, half the width of the building. The pitch of a roof is the ratio of the total rise of the roof to the total width of the building.

In determining rafter size, the *rise* is considered to be the vertical distance from the top of the wall plate to the upper end of the measuring line. To find the rise of a roof, multiply the pitch by the span.

> **Illustration.** A building is 24′ wide and has a roof with a 1/4 pitch, figure 8-1. What is the rise?
>
> **Solution.**   24′ x 1/4 = 24/4 = 6′
>
> **Explanation.** • In this illustration, the solution involves multiplication of a fraction by a whole number: 24′ x 1/4 = 6′.

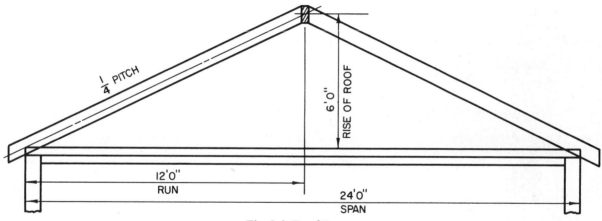

Fig. 8-1 Roof Layout

Determining Board Feet

Lumber used in construction may be measured in board feet. The board foot measurement is 1″ thick, 1′ wide, and 1′ long, figure 8-2.

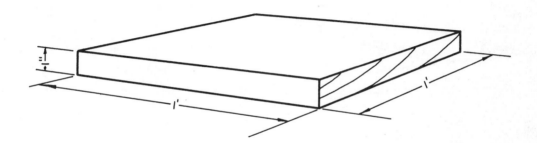

Fig. 8-2 Board Foot

Rough lumber measures to standard (nominal) thicknesses of 3/8'', 1/2'', 5/8'' and 1''. Lumber 1'' or less in thickness is figured as if it were 1'' thick. For lumber that is more than 1'' thick, the nominal stock size is used.

Note: *Nominal* is a term referring to the commercial size given to lumber or other material, rather than the actual size.

The width of boards is figured in even widths with the exception of 3'' and 5'' boards. Other uneven dimensions are expressed as the next larger even width. The nominal widths of lumber are 2'', 3'', 4'', 5'', 6'', 8'', 10'', and 12''. A 5 1/4- or 5 1/2-inch board is figured as 6'' wide and a 7 1/4- or 7 1/2-inch board is figured as 8'' wide.

There are three formulas used to determine the size of lumber, a basic formula and two others which are derived from the first.

Note: A *formula* is a short method of expressing a relationship in arithmetic by the use of symbols, which are in the form of letters. Known values may be then substituted for the symbols.

$$\text{Number of pieces } \times t'' \times w' \times 1' = \text{board feet}$$

(t = thickness; w = width; 1 = length).

Illustration. How many board feet are there in 2 pieces of 3/8'' x 2' x 8' plywood?

Solution. Express 3/8'' as 1''.

2 x 1'' x 2' x 8' = 32 board feet (bd. ft.)

Explanation. • After changing 3/8'' to 1'', proceed with the multiplication.

When the width or length is expressed in inches, figure 8-3, use the formula:

$$\frac{\text{Number of pieces } \times t'' \times w'' \times 1'}{12}$$

Illustration. Find the number of board feet in 5 pieces of lumber which measure 3/4'' x 7 1/2'' x 12'.

Solution. Express 3/4'' as 1''. Express 7 1/2'' as 8''.

$$\frac{5 \times 1'' \times 8'' \times 12'}{12} = \frac{5 \times 1'' \times 8'' \times \overset{1}{\cancel{12'}}}{12} = 40 \text{ bd. ft.}$$

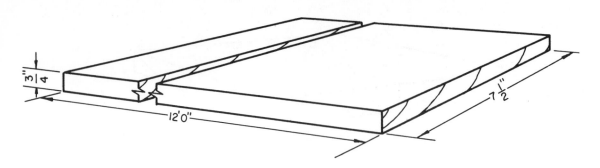

Fig. 8-3 Board Foot (width expressed in inches)

Explanation. • If either the width or length is expressed in inches, it must be converted to feet by dividing by 12. (In the formula, 12 is placed in the denominator). • The 3/4″ and 7 1/2″ are changed to whole-inch measurements. • Use the cancellation process and then multiply for the result.

If the thickness, width and length are all expressed in inches, figure 8-4, use the formula:

$$\frac{* \text{ Number of pieces x t'' x w'' x 1''}}{144}$$

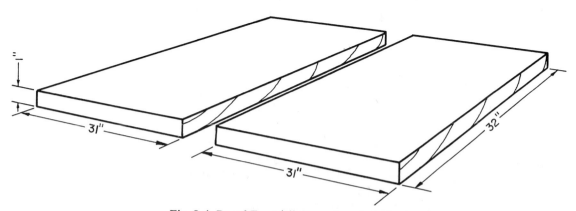

Fig. 8-4 Board Foot (all dimensions in inches)

Illustration. Find the number of board feet in 2 pieces of lumber which measure 1″ x 32″ x 31″.

Solution.

$$\frac{\text{Number of pieces} \ \text{x} \ \text{t} \ \text{x} \ \text{w} \ \text{x} \ 1}{144} = \frac{2 \text{ x } 1'' \text{ x } 32'' \text{ x } 31''}{144} = \frac{124}{9} = 13 \ 7/9 \text{ bd. ft.}$$

Explanation. • If both the width and length are expressed in inches, each must be converted to feet by dividing by 12. (In the formula, 12 x 12, or 144, is placed in the denominator.) • Reduce the size of the numbers. • Multiply, and change the improper fraction to a whole or mixed number by dividing 124 by 9.

When the lumber is more than 1″ thick, figure 8-5, page 46, use the nominal measurement.

Illustration. Find the number of board feet in 2 pieces of door jamb stock which measure 1 1/4″ x 5 3/4″ x 8′.

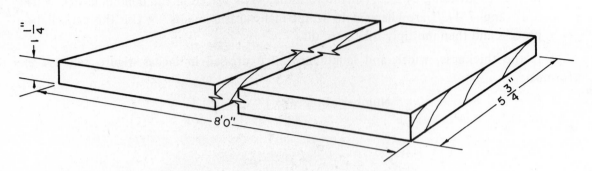

Fig. 8-5 Board Foot (more than 1″ thick)

Solution.   Change 1 1/4″ to 5/4″. (4 x 1 + 1 = 5/4)

Express 5 3/4″ as 6″.

$$\frac{2 \times 5'' \times \overset{1}{\cancel{6}}'' \times \overset{1}{\cancel{8}'}}{\underset{1}{\cancel{4}} \times \underset{2}{\cancel{12}}} = \frac{10}{1} = 10 \text{ bd. ft.}$$

Explanation. • After the thickness (1 1/4) is changed to an improper fraction and the width (5 3/4) is expressed as the next highest even number, multiply as in preceding illustrations using 12 in the denominator.

• To change a mixed number to an improper fraction, multiply the whole number (1) by the denominator (4) and add the numerator (1).

COMBINING ADDITION, DIVISION AND MULTIPLICATION

To calculate the size of the boards when they vary in width, determine the average width by the process of addition and division; then multiply using the correct formula.

Illustration. Find the number of board feet in the piece shown in figure 8-6. The piece is 1″ thick; 10″, 6″ and 8″ wide; and 12′ long.

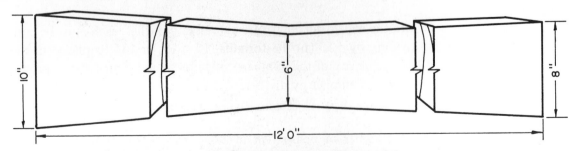

Fig. 8-6 Board Foot (with varying widths)

Solution. $10'' + 6'' + 8'' \div 3 = 8''$ wide

$$\frac{1'' \times 8'' \times \overset{1}{\cancel{12}}}{\underset{1}{\cancel{12}}} = 8 \text{ bd. ft.}$$

Explanation. • To find the average width add all given widths $(10 + 6 + 8 = 24)$. • Then divide by the number of measurements stated as widths (3). • The quotient $(8'')$ is used as the width of the piece.

PRACTICAL APPLICATION

The size of a footing for a building can be determined by multiplication. The *footing* of a building is a concrete base upon which the foundation walls and piers are built to distribute the weight of the superstructure.

Illustration.   Determine the footing size for an average sized one-story building with $8''$ walls.

Solution. $8'' \times 2 = 16''$ width of footing

   ($8''$ thickness same as width of the wall)

   $8'' \times 1/2 = 4''$ projection on each side of wall

Explanation. • The width of the footing is twice the width of the wall. • The thickness is the same as the width of the wall, and the projection is one-half times the width of the wall.

CANCELLATION

A process which can be used when multiplying fractions is known as *cancellation*. In cancellation, the numerator of one fraction and the denominator of the other fraction are divided by the same number. It is important to remember that this process may be used only in the multiplication of fractions; not when adding or subtracting fractions.

Illustration. Multiply $25/6 \times 36/5$, using the cancellation process.

Solution.

$$\frac{\overset{5}{\cancel{25}}}{6} \times \frac{36}{\underset{1}{\cancel{5}}} = \text{(Divide by 5.)}$$

$$\frac{5}{\underset{1}{\cancel{6}}} \times \frac{\overset{6}{\cancel{36}}}{1} = \text{(Divide by 6.)}$$

$$\frac{30}{1} = 30$$

Explanation. • Both 25 and 5 are cancelled by 5. • Both 36 and 6 are cancelled by 6. • The answer is 30/1 or 30.

APPLICATION

1. Multiply the following fractions. Reduce them to their lowest terms if necessary.

   a. 3/4 x 1/2   c. 2/3 x 2/5   e. 29/32 x 33/64
   b. 3/8 x 5/8   d. 3/7 x 3/14   f. 7/38 x 9/45

2. Multiply the following fractions.

   a. 5/8 x 1/3 x 1/4   c. 3/4 x 1/2 x 3/7   e. 3/64 x 1/2 x 1/3
   b. 2/3 x 5/8 x 1/2   d. 7/8 x 4/35 x 1/10   f. 7/25 x 3/32 x 1/2

3. If 2/3 of an hour is considered the average time for installation of a ceiling light such as the one shown below, how many ceiling lights can be installed in 8 hours at the same rate?

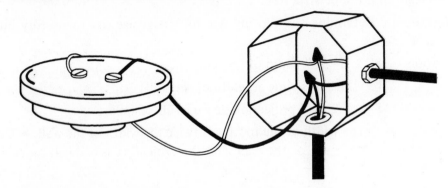

4. A pipe is cut one quarter the length of a pipe 7/8″ long. How long is the small piece?

5. How many board feet are there in the following pieces of lumber?

   a. 2 pcs., 1/2″ x 6″ x 12′   d. 3 pcs., 2″ x 5″ x 12′
   b. 5 pcs., 1″ x 4″ x 10′   e. 2 pcs., 4″ x 3″ x 12′
   c. 2 pcs., 1 1/4″ x 6″ x 8′   f. 2 pcs., 3/4″ x 8″, 6″, 4″ and 10′

6. A truck carries 1/5 ton of concrete in one load. What is the total amount that a driver delivers in 5, 8, 10 and 12 trips?

7. A bricklayer lays a wall 3/4 foot high in one hour. If he continues at the same rate, how high will the wall be in 8 hours?

8. Find the products of the following whole numbers and fractions.

   a. 6 x 1/3   e. 8 x 2/3   i. 9 x 4/7
   b. 5 x 3/5   f. 7 x 3/4   j. 12 x 2/3
   c. 3 x 5/6   g. 8 x 7/12   k. 12 x 3/16
   d. 5 x 5/9   h. 7 x 4/5   l. 10 x 5/8

9. A building is 24′ wide. The roof has a 1/4 pitch. What is the rise?

10. A building is 32′ wide. The roof has a 1/3 pitch. What is the rise?

Unit 9
Division of Fractions

OBJECTIVES

After studying this unit, the student should be able to

- divide a fraction by another fraction.

- divide a whole number by a fraction.

- find the unknown factor when one factor and the product are known.

The process of division is similar to multiplication of fractions except that in the division of fractions, the terms of the divisor are reversed, or *inverted.*

DIVIDING A FRACTION BY ANOTHER FRACTION

To divide a fraction by a fraction, invert the terms of the divisor and multiply the numerator by the numerator and the denominator by the denominator. All mixed numbers should be reduced to improper fractions when they are to be divided.

Illustration. How many times is 9/64 contained in 3/16?

Solution.

$$\frac{3}{16} \div \frac{9}{64} = \frac{\cancel{3}^{1}}{\cancel{16}_{\cancel{4}_1}} \times \frac{\cancel{64}^{\cancel{16}^4}}{\cancel{9}_3} = \frac{4}{3} = 1\frac{1}{3}$$

Explanation. The divisor is 9/64. • Invert the divisor, which means to change the position of the numerator and denominator. The problem then becomes a multiplication process. • The division sign (÷) is changed to a multiplication sign (x). • Cancellation is used to reduce the size of the numbers to be multiplied.

DIVIDING A FRACTION BY A WHOLE NUMBER AND
A WHOLE NUMBER BY A FRACTION

All whole numbers have 1 as a denominator; therefore, they can be treated as a fraction.

Illustration. Studs are placed 16″ O.C. for a wall 48′ long. If one stud is added as a starter, how many studs are there?

Solution. 48′ ÷ 16″

Step 1. Change 16″ to feet = 1 1/3′.

Step 2. Change 1 1/3 to an improper fraction (4/3).

Step 3.

$$48 \div \frac{4}{3} = \overset{12}{\cancel{48}} \times \frac{3}{\underset{1}{\cancel{4}}} = \frac{36}{1} = 36$$

Step 4. 36 + 1 = 37

Explanation. Notice that 16″ is changed to feet because division requires that dimensions be expressed in like units. The mixed number is changed to an improper fraction. The denominator of 48 (1) is written. The divisor is inverted. The result is obtained by multiplication. One stud is added for a starter.

A similar procedure is followed when the divisor is a whole number.

Illustration. Divide 1/2 by 4.

Solution.   1/2 ÷ 4 = 1/2 ÷ 4/1 = 1/2 x 1/4 = 1/8

Explanation. • Remember that the divisor is the fraction which is inverted.

COMBINING MULTIPLICATION AND DIVISION OF FRACTIONS

At times the carpenter must solve problems requiring multiplication and division.

Illustration. (1/2 x 4) ÷ (1/4 ÷ 2)

Solution.   Step 1. 1/2 x 4/1 = 4/2 = 2
                 Step 2. 1/4 ÷ 2/1 = 1/4 x 1/2 = 1/8
                 Step 3. 2 ÷ 1/8 = 2 x 8/1 = 16

Explanation. • The parentheses in the original problem indicate that the processes inside them are performed first. • The product of 1/2 x 4 (2) becomes the dividend and the quotient of 1/4 ÷ 2 (1/8) becomes the divisor. • The answer is 16.

APPLICATION

1. Find the quotient for each of the following pairs of proper fractions.

   a. 5/8 ÷ 1/2   h. 7/32 ÷ 7/8   o. 13/16 ÷ 5/16
   b. 3/8 ÷ 1/32   i. 17/32 ÷ 23/32   p. 3/8 ÷ 13/16
   c. 2/3 ÷ 7/16   j. 5/16 ÷ 11/16   q. 9/16 ÷ 5/32
   d. 9/16 ÷ 3/8   k. 1/2 ÷ 3/8   r. 7/32 ÷ 2/3
   e. 13/16 ÷ 3/32   l. 3/8 ÷ 5/32   s. 3/8 ÷ 17/32
   f. 3/8 ÷ 15/32   m. 2/3 ÷ 7/8   t. 5/16 ÷ 5/8
   g. 5/32 ÷ 15/16   n. 9/16 ÷ 17/32

2. The centers of 4 dowel holes are spaced equal distances apart and from the ends. The stock is 27 1/2 inches long. What is the length of each measurement?

3. Divide each whole number by the common fraction in the following.

   a. 12 ÷ 1/2   i. 12 ÷ 3/4   p. 50 ÷ 1/10
   b. 10 ÷ 1/4   j. 10 ÷ 1/2   q. 16 ÷ 1/2
   c. 19 ÷ 1/3   k. 21 ÷ 3/8   r. 2 ÷ 1/3
   d. 6 ÷ 3/4   l. 9 ÷ 4/9   s. 17 ÷ 3/10
   e. 11 ÷ 5/9   m. 18 ÷ 3/6   t. 56 ÷ 3/7
   f. 11 ÷ 1/6   n. 15 ÷ 1/4   u. 64 ÷ 3/8
   g. 8 ÷ 4/9   o. 40 ÷ 5/8   v. 13 ÷ 7/9
   h. 91 ÷ 1/2

4. Divide each fraction by the whole number in the following.

   a. 4/5 ÷ 9   k. 1/3 ÷ 19   t. 1/4 ÷ 15
   b. 5/16 ÷ 64   l. 4/9 ÷ 8   u. 1/10 ÷ 50
   c. 3/4 ÷ 4   m. 3/5 ÷ 4   v. 9 ÷ 4/9
   d. 5/9 ÷ 11   n. 10 ÷ 2/5   w. 3/4 ÷ 16
   e. 1/2 ÷ 10   o. 1/2 ÷ 12   x. 7/9 ÷ 13
   f. 3/8 ÷ 21   p. 1/6 ÷ 11   y. 17 ÷ 3/10
   g. 3/6 ÷ 18   q. 1/2 ÷ 91   z. 56 ÷ 3/7
   h. 1/3 ÷ 2   r. 6 ÷ 3/4   aa. 1/4 ÷ 10
   i. 5/8 ÷ 40   s. 3/4 ÷ 12   ab. 5/6 ÷ 9
   j. 3/8 ÷ 64

5. A cabinet 35″ high is designed to contain shelves 3/4″ thick spaced 8″ apart. How many shelves will it hold?

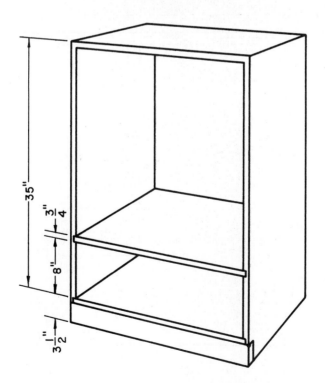

6. Find the unknown number in each of the following.

a. 2/3 x _ = 6   g. 4/5 x _ = 4   m. 1/4 x _ = 6
b. 3/8 x _ = 16   h. 3/16 x _ = 20   n. 1/3 x _ = 9
c. 3/7 x _ = 14   i. 5/8 x _ = 12   o. 6/7 x _ = 7
d. 5/6 x _ = 10   j. 3/4 x _ = 10   p. 9/10 x _ = 20
e. 7/8 x _ = 6   k. 1/2 x _ = 8   q. 2/9 x _ = 18
f. 3/16 x _ = 9   l. 7/12 x _ = 14   r. 3/5 x _ = 8

7. Find the unknown number in each of the following.

a. _ x 1/4 = 8   g. _ x 5/8 = 14   m. _ x 1/3 = 9
b. _ x 6/7 = 14   h. _ x 5/16 = 15   n. _ x 2/3 = 7
c. _ x 9/10 = 15   i. _ x 3/16 = 12   o. _ x 2/9 = 10
d. _ x 7/12 = 16   j. _ x 4/5 = 8   p. _ x 3/8 = 4
e. _ x 1/2 = 10   k. _ x 2/3 = 9   q. _ x 3/4 = 16
f. _ x 3/11 = 10   l. _ x 3/8 = 12   r. _ x 9/16 = 32

8. Solve the following problems by multiplication and division.

a. (3/4 x 4) ÷ (1/2 ÷ 2) =   c. (1/2 x 7) x (3/8 ÷ 1/2) x 2 =
b. (3/4 x 8) ÷ (5/8 ÷ 4) =   d. [(4/5 ÷ 1/3) x (4 x 1/4)] ÷ 1/2 =

9. If striking a nail once drives it 1/4″ into a board, how many times must it be hit to drive it 3/4″?

10. The scale on the drawing below is 1/4″ = 1′. How many feet are represented in a distance which measures 3/4″?

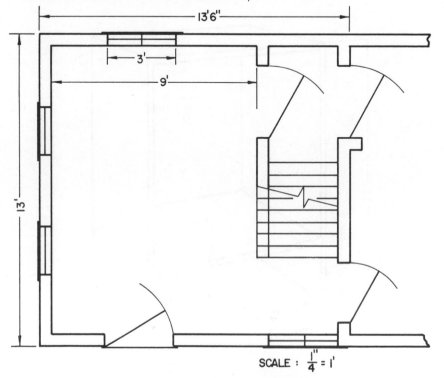

SCALE : 1/4″ = 1′

11. How many pieces of wire 3/4" long can be cut from a piece of wire 3' long? (Change feet to inches.)

12. How many pieces 3/4" long can be cut from the 15" long pipe in the figure?

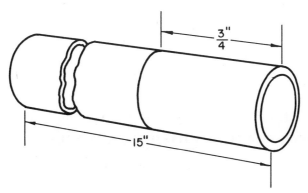

13. The scale on the blueprint of a commercial building is 1/8" = 1'. Find the actual dimension of an area measuring 6 inches long.

14. How many lengths of pipe 6" long can be cut from a pipe 20' long?

15. How many studs are required for a wall 64' long when the studs are spaced 16" O.C.?

16. How many rafters are required for a roof 64' long when the rafters are spaced 18" O.C.?

Unit 10
Addition of Mixed Numbers

OBJECTIVES

After studying this unit, the student should be able to

- change improper fractions to equivalent mixed numbers.
- reduce improper fractions to their lowest terms.
- add two or more mixed numbers.

CHANGING IMPROPER FRACTIONS TO MIXED NUMBERS

The horizontal line between the numerator and denominator of a fraction indicates division, just as the problem $4 \div 3$ indicates division of whole numbers.

To change an improper fraction to a mixed number, divide the numerator of the fraction by its denominator: $5/3 = 1\ 2/3$.

Note: When the numerator is the same as the denominator, the fraction is equivalent to a whole number: $3/3 = 1$.

Illustration. The length of a foundation wall for a house is expressed as 206/3'. Reduce the improper fraction to a mixed number.

Solution.

$$\frac{206'}{3} = 3 \overline{)\begin{array}{l} 68\ 2/3' \\ 206' \end{array}}$$

$$\begin{array}{r} \underline{18} \\ 26 \\ \underline{24} \\ \underline{2} \\ 3 \end{array}$$

Explanation. • Follow the procedure for changing an improper fraction to a mixed number: divide the numerator (206) by the denominator (3). The quotient (68) is the whole number of the mixed number. • The remainder (2) is placed as a numerator over the original denominator. • Write the whole number and the fraction as a mixed number (68 2/3).

To prove that 68 2/3 is the correct answer, apply the procedure for changing a mixed number to an improper fraction. The result is the original number (206/3):

$$68\frac{2}{3} = \frac{68 \times 3 + 2}{3} = \frac{206}{3}$$

REDUCING AN IMPROPER FRACTION TO LOWEST TERMS

An improper fraction can be reduced to its lowest terms before being changed to a mixed number. When several fractions are grouped for addition, reducing them before changing to improper fractions helps to simplify the problem.

Illustration. Change 50/16 to its lowest terms.

Solution. 25

$$\frac{\cancel{50}}{\cancel{16}} = \frac{25}{8}$$

8

Explanation. • Reduce by cancellation.

ADDING TWO OR MORE MIXED NUMBERS

To add two or more mixed numbers, add the whole numbers and then add the fractions. When the sum of the fractions is an improper fraction, change it to a mixed number. Add the whole number to the other whole numbers and annex the fraction.

Illustration. The layout on a wall for a duplex outlet is 3 1/2', 2 1/4', and 5 3/8', figure 10-1. What is the total distance?

Solution.   3 1/2' + 2 1/4' + 5 3/8'

Step 1. 3 + 2 + 5 = 10

Step 2. Least common denominator is 8. 1/2 = 4/8; 1/4 = 2/8; 3/8 = 3/8

Step 3. 4/8 + 2/8 + 3/8 = 9/8   9/8 = 1 1/8

Step 4. 10 + 1 1/8 = 11 1/8

3 1/2 + 2 1/4 + 5 3/8 = 3 4/8 + 2 2/8 + 5 3/8 = 11 1/8'

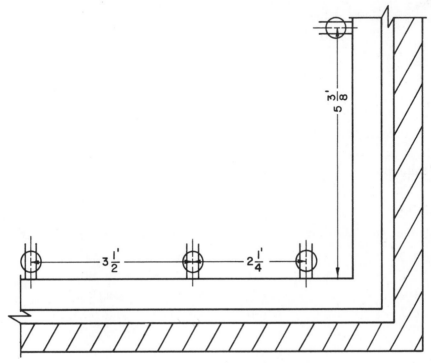

Fig. 10-1 Duplex Outlet Layout

Explanation. • Add the whole numbers (3, 2, and 5) and write the sum (10). • Reduce fractions to their least common denominators (4/8, 2/8, and 3/8). • Add the fractions and reduce the sum to a mixed number. • Add the sum of the fractions (1 1/8) to the sum of the whole (10); the result is 11 1/8.

Whole numbers, mixed numbers, and improper fractions may be added together.

Illustration. 5 + 14 1/2 + 13 1/4 + 21/8

Solution.   Step 1. 21/8 = 2 5/8

Step 2. 5 + 14 + 13 + 2 = 34

Step 3. 1/2 = 4/8; 1/4 = 2/8; 5/8 = 5/8

Step 4. 4/8 + 2/8 + 5/8 = 11/8 = 1 3/8

Step 5. 34 + 1 + 3/8 = 35 3/8

Explanation. • Reduce the improper fraction (21/8) to a mixed number (2 5/8). • Add the whole numbers (5 + 14 + 13 + 2 = 34). After finding the L.C.D. (8), reduce each fraction to lowest terms (4/8, 2/8, and 5/8). • Add the fractions. • Change the improper fraction (11/8) to a mixed number (1 3/8). • Add the 1 to the whole number (34) and annex the fraction. • The sum is 35 3/8.

ADDING MIXED NUMBERS IN A VERTICAL COLUMN

After sufficient practice, it is possible to add several mixed numbers in a vertical column. This can be done with accuracy and speed by combining steps. The procedures of the addition are the same as those already illustrated in this unit.

Illustration. Find the sum of 4 3/4, 3 1/2, 1 1/16, and 2 3/32.

Solution.   4 3/4 = 4 24/32

3 1/2 = 3 16/32

1 1/16 = 1 2/32

2 3/32 = 2 3/32

10 45/32 = 11 13/32

Explanation. • Write the mixed numbers in a vertical column and add the whole numbers. • Reduce the fractions to the least common denominator (32). • Add the fractions and reduce the sum to a whole or mixed number. • Add the result to the sum of the whole numbers.

APPLICATION

1. Reduce the following improper fractions to a whole or mixed number.

a. 4/3	h. 8/4	o. 16/5	u. 8/4	aa. 80/40
b. 5/2	i. 10/3	p. 4/1	v. 75/25	ab. 90/45
c. 10/5	j. 7/6	q. 32/7	w. 8/5	ac. 98/22
d. 6/3	k. 23/8	r. 64/13	x. 75/10	ad. 125/25
e. 8/3	l. 50/10	s. 16/1	y. 150/100	ae. 108/12
f. 10/6	m. 33/14	t. 6/5	z. 78/56	af. 160/16
g. 7/4	n. 8/5			

2. Reduce the following fractions to lowest terms and change to whole or mixed numbers.

    a. 60/50   f. 90/45   k. 14/5   p. 33/33   u. 47/13
    b. 64/60   g. 90/84   l. 15/13   q. 75/25   v. 100/100
    c. 40/32   h. 64/16   m. 27/26   r. 80/20   w. 41/3
    d. 100/46   i. 8/5   n. 51/8   s. 60/30   x. 34/33
    e. 48/28   j. 13/3   o. 120/120   t. 50/25

3. Reduce the following fractions to their least common denominators, and add.

    a. 3/4, 1/2, 1/16   e. 3/64, 3/32, 5/8
    b. 3/4, 3/8, 3/16   f. 9/32, 15/16, 2/9
    c. 5/8, 3/5, 3/10   g. 5/8, 3/10, 13/16, 4/39
    d. 2/3, 3/4, 2/9   h. 4/15, 3/13, 2/27

4. Add the following mixed numbers.

    a. 3 5/8 + 2 7/12 + 3 11/24   g. 7 7/8 + 5 2/3 + 6 1/4
    b. 2 5/8 + 3/7 + 3/5   h. 13/16 + 5/24 + 11/32
    c. 2 3/20 + 1 7/12 + 5 5/15   i. 2 5/12 + 1 1/4 + 11 5/24
    d. 4 1/3 + 3 2/7 + 1/4   j. 3 1/2 + 2 1/3 + 5 3/4
    e. 3 2/7 + 3 2/3 + 3 4/15   k. 5 + 1/2 + 6 1/2 + 3 1/3
    f. 7/8 + 5/18 + 8/9   l. 5/9 + 4/5 + 1/2

5. Add the following combination of numbers.

    a. 2 + 12 1/3 + 2 1/6   e. 3/2 + 4/1 + 2 + 2 3/4
    b. 1/15 + 15/3 + 5   f. 2 1/2 + 47/3 + 5 + 5/8
    c. 5/2 + 6/5 + 1 1/5 + 3   g. 2/9 + 9/2 + 35/12 + 1/4
    d. 39/6 + 47/3 + 5 + 5/8   h. 2/2 + 4/4 + 4/3 + 3/2

6. The foundation sill shown below is marked in lengths of 2 1/2', 2 3/4', 2', 2 2/3', 3 3/4', and 3 1/4'. How long is the sill? *Note:* A *foundation sill* is a plank or timber which rests upon a foundation wall.

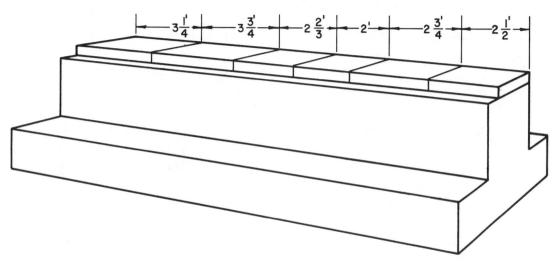

7. Determine the width and length of the window frame (in inches) in the figure below.

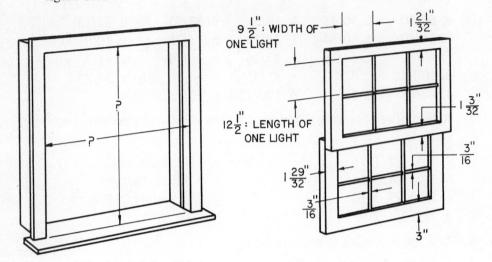

8. Determine the width and length of the door jamb (in inches) in the figure below.

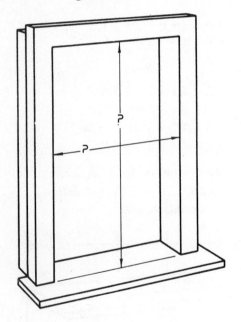

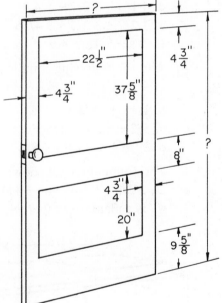

9. A plumber receives lengths of 4″ cast-iron pipe which measure 4 1/4′, 2 1/2′, 3 1/8′, and 3 3/4′ and lengths of 2″ cast-iron pipe which measure 1 1/2′, 3 3/8′, 4 1/4′, and 4 3/4′. How many feet of each size pipe are received?

10. An electrical job requires 1 1/2 rolls of #6 wire, 2 1/8 rolls of #12 wire, and 1 5/12 rolls of #14 wire. One roll equals 250′. How many feet of wire are required for the job?

11. Three bathroom fixtures weigh 195 3/8 lb., 20 1/2 lb., and 45 5/16 lb. What is their total weight?

12. A plumber works 2 1/2 hours on one job, 2 3/8 hours on another, and 5 1/4 hours on a third job. What is the total amount of time worked?

13. A sheet metal worker has three pieces of pipe which measure 2 5/16″, 2 3/32″, and 2 3/4″. What is the total length?

14. What is the outside diameter of the pipe in the figure below?

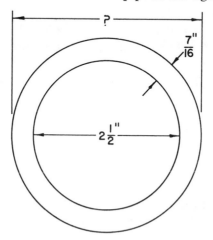

15. A bricklayer requires 2 1/2 days to do one job, 3 1/4 days to do another job, and 8 1/10 days to do a third job. How many days are needed to complete all three jobs?

Unit 11
Subtraction of Mixed Numbers

OBJECTIVES

After studying this unit, the student should be able to

- subtract mixed numbers.
- solve problems that involve addition and subtraction of mixed numbers.

In the subtraction of mixed numbers, the fractions are subtracted first, and then the whole numbers are subtracted. The remainders of the fractions and whole numbers are combined to form one number.

In the subtraction of whole numbers, the minuend must be larger than the subtrahend. Sometimes it becomes necessary to borrow units of mixed numbers.

Illustration. A piece of pipe elbow 4 3/4″ long is cut from the original pipe which measures 9 1/3″, figure 11-1. How many inches remain?

Solution. L.C.D. = 12
$$9\ 1/3 = 9\ 4/12 = 8\ 16/12$$
$$4\ 3/4 = 4\ 9/12 = \underline{4\ \ \ 9/12}$$
$$4\ \ 7/12$$

Explanation. • Find the least common denominator (12). In this case, 9/12 cannot be subtracted from 4/12 since the minuend must be larger than the subtrahend before subtraction can take place. • If a minuend in a fraction is too small for subtraction, take 1 unit from the whole number, change it to the equivalent fraction, and add it to the numerator of the fraction. • Take 1, or 12/12, from 9 which leaves 8. • Add 12/12 to 4/12 which gives 16/12 and subtract 4 9/12 from 8 16/12. • Subtract the fractions (16/12 – 9/12) and the whole numbers (8 – 4). • The answer is 4 7/12.

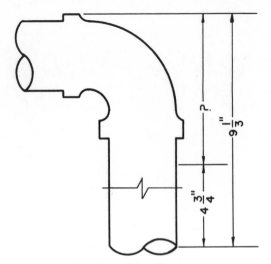

Fig. 11-1 Pipe Elbow

60

SUBTRACTING A FRACTION FROM A MIXED NUMBER

When solving problems which require subtracting a fraction from a mixed number, first subtract the fraction from the fractional part of the mixed number. Annex this fractional difference to the whole number. The end result may be a mixed fraction or a whole number.

Illustration. The hot water pipe shown in figure 11-2 is 3 7/8″ from the vent pipe. The cold water pipe is 9/16″ closer to the vent pipe. How far is the cold water pipe from the vent pipe?

Solution.   3 7/8   = 3 14/16
                – 9/16 =   9/16
                            3 5/16

Explanation. • Arrange the numbers in a vertical column placing the minuend (3 7/8″) above the subtrahend (9/16″). • Reduce the fractions to least common denominators. • Subtract 9/16″ from 14/16″, leaving 5/16″. Write the whole number (3) and annex the fraction (5/16″). The answer is 3 5/16″.

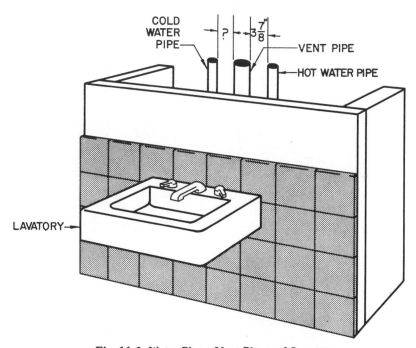

Fig. 11-2 Water Pipes, Vent Pipe and Lavatory

SUBTRACTING A MIXED NUMBER FROM A WHOLE NUMBER

To subtract a mixed number from a whole number, borrow 1 unit from the whole number and change it to a fraction. Subtract the fractions and then the whole numbers.

Illustration. A carpenter saws 2 1/2″ from a 4″ block. How much is left of the original block?

Solution.   4″ – 2 1/2″   4   = 3 2/2
                            – 2 1/2 = 2 1/2
                                        1 1/2

Explanation. • Write the numbers in a vertical column. • Borrow 1 unit from the minuend, leaving 3 as the whole number. • Change the 1 to a fraction (2/2) which has the same denominator as the fraction in the subtrahend. • Subtract the fraction 1/2 from 2/2, leaving 1/2 and then subtract the whole numbers. • The answer is 1 1/2″.

Finding The Difference in Level of Two Points

To be able to find the difference in level of two points is essential in laying out foundations. The instrument in figure 11-3 consists of a tripod which supports a small telescope with crossed hair wires within. The observer may adjust the line of sight to determine the level of two points. The figures are subtracted to find the difference in level.

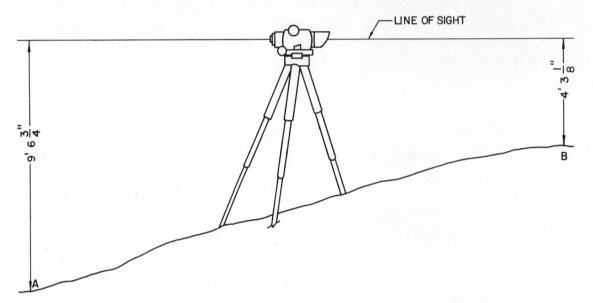

Fig. 11-3 Leveling Instrument

Illustration.   What is the difference in level headings of 4′ 3 1/8″ and 9′ 6 3/4″?

Solution.   9′ 6 3/4″ = 9′ 6 6/8″
            − 4′ 3 1/8″ = 4′ 3 1/8″
                           5′ 3 5/8″

Explanation. • In figure 11-3, the reading on the target rod at A is 9′ 6 3/4″ and at B is 4′ 3 1/8″. • Subtract the target reading at B from the target reading at A. The difference in height from A to B is 5′ 3 5/8″.

COMBINING ADDITION AND SUBTRACTION OF MIXED NUMBERS

In determining certain measurements, addition and subtraction of mixed numbers are sometimes required.

Illustration. A pipe measures 17 1/2′ long. A piece 3 1/2′ long and a piece 5 1/4′ long are cut off the pipe. How much is left of the original pipe?

Solution.   17 1/2' – (3 1/2' + 5 1/4') = 8 3/4'

Add: 3 1/2 = 3 2/4	Subtract: 17 2/4 = 16 6/4
+ 5 1/4 = 5 1/4	- 8 3/4 = 8 3/4
8 3/4	8 3/4

Explanation. • The fractions should have common denominators. • Add the lengths cut off (3 2/4' + 5 1/4'). The sum is 8 3/4'. • Subtract 8 3/4' from 17 2/4'. • Since the subtrahend fraction is larger than the minuend fraction, borrow 1 unit from 17, leaving 16, and add the 1 to 2/4, making the fraction 1 2/4. • Change 1 2/4 to an improper fraction, 6/4. • Subtract the fractions and then subtract the whole numbers. Combine the whole number and the fraction. There are 8 3/4' left on the original pipe.

APPLICATION

1. A board is 16 1/2" long. What length board is needed which, when combined with the 16 1/2" board, fills a space 30 3/4" long?

2. After a 2 3/4' piece is sawed from a 14 1/2' board, how much is left?

3. After 13 5/8" is cut from a pipe 28" long, how much is left?

4. Rough lumber is thicker than dressed lumber. A rough board measuring 2" thick is dressed by removing 3/16" off one surface and 1/4" off the other surface. How thick is the dressed board?

5. A carpenter purchases 950 7/10 bd. ft. of sheathing. He uses 505 3/4 bd. ft. How much is left?

6. A carpenter has 75 3/4 lb. of #8d common nails and uses 41 1/8 lb. How many pounds are left?

7. From the sum of 3 1/3 and 6 1/4 subtract the sum of 9/16 and 4 1/4.

8. Subtract the following.

   a. 2 1/2 from 5   f. 10 7/12 from 14
   b. 5 3/4 from 12 3/8   g. 21 9/16 from 31 13/32
   c. 3 4/5 from 7 7/8   h. 59/2 from 84/2
   d. 1 7/10 from 4 1/3   i. 3 1/2 from 80/5
   e. 19 1/6 from 42 3/8   j. 16/4 from 8 1/8

9. Find the difference between 7 1/3 and 6 5/6.

10. Subtract the following pairs of mixed numbers.

5 1/5	6 3/16	5 1/2	6 4/9	8 7/16	9 5/12
- 2 3/8	- 4 3/32	- 3 7/8	- 3 1/3	- 3 3/64	- 5 3/8

7 3/7	5 7/8	10 7/12	12 1/2	5 3/8	8 4/5
- 2 1/3	- 2 3/5	- 6 2/3	- 2 2/3	- 2 1/4	- 1 2/15

11. The wall thickness of the metal pipe is 1/8″ and the outside diameter is 3 1/2″. Find the inside diameter of the pipe.

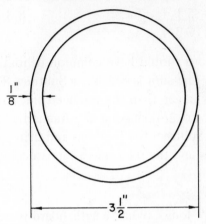

12. How many inches of bricks are left on a wall measuring 87 1/2″ high after 15 3/8″ are taken off?

13. A plumber worked 1 1/2 hours installing a hot water tank, 2 1/4 hours repairing a space heater, and 2 5/6 hours repairing a stove. How much time is left from an 8-hour workday?

14. A mason finishes 6 5/8 cu. yd. of a concrete slab which requires 10 2/3 cu. yd. for the complete job. How much remains to be finished?

15. One electrician can install a ceiling light and a wall switch in 4 3/8 hours, while another performs the same job in 6 5/6 hours. How much less time does it take the first electrician to do the job than the second?

16. A structure requires a pipe 11/16″ long. The pipe cut for the job is 41/64″ long. How much longer should the pipe have been cut?

17. The wall thickness of a piece of tubing is 7/16″; its outside diameter is 3 3/4″. Find the inside diameter.

18. Find the difference in weight between 11 3/4 pounds of #6d finishing nails and 6 3/5 pounds of #8d common nails.

19. A carpenter uses pieces of copper tubing measuring 4 1/2′ and 2 1/4′. What is the difference in length?

20. A carpenter buys 5000 square feet of asphalt shingles. On one building, 285 1/2 square feet and 325 3/8 square feet are used; on another building, 2109 5/16 square feet and 1055 3/16 square feet are used. How many square feet of shingles remain?

21. A carpenter agrees to complete a job in 60 hours. He works 7 3/4, 9 1/2, 22 3/8, and 15 9/16 hours. How many hours remain out of the 60 hours?

22. Find the difference in level of target readings which are 2′ 6 1/2″ and 6′ 5 3/4″.

Unit 12
Multiplication of Mixed Numbers

OBJECTIVES

After studying this unit, the student should be able to

- multiply mixed numbers by whole numbers, mixed numbers by fractions, and mixed numbers by mixed numbers.

- solve problems requiring multiplication of mixed numbers combined with other processes.

Multiplication of mixed numbers is used frequently in all areas of construction.

MULTIPLYING MIXED NUMBERS BY WHOLE NUMBERS

In the process of multiplying a mixed number by a whole number, it is important to remember that 1 is the denominator of the whole number and that the mixed number must be changed to an improper fraction. The numerators are then multiplied by the numerators to form a new numerator, and denominators are multiplied by denominators to form a new denominator.

Illustration. The space for a switch box outlet as shown in figure 12-1 measures 2 1/4″ wide. How much space is required for 5 outlets?

Solution.   5 x 2 1/4″ = 5/1 x 2 1/4 = 5/1 x 9/4 = 45/4 = 11 1/4″

Explanation. • The denominator of the whole number 5 is 1 (5/1); 2 1/4 is changed to an improper fraction (9/4). • The numerator (5) is multiplied by the numerator (9). • The denominator (1) is multiplied by the denominator (4). • The product (45) is the numerator of the improper fraction and the product (4) is the denominator. • Reduce the improper fraction (45/4) to a whole or mixed number. The result is 11 1/2″.

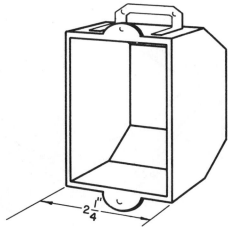

Fig. 12-1 Switch Box

MULTIPLYING MIXED NUMBERS BY FRACTIONS

To multiply a mixed number by a fraction, reduce the mixed number to an improper fraction. The multiplication is carried out in the same manner as when common fractions are multiplied.

Illustration. Multiply 3/8 by 2 3/4.

Solution. 3/8 x 2 3/4 = 3/8 x 11/4 = 33/32 = 1 1/32

Explanation. • Reduce the mixed number (2 3/4) to an improper fraction (11/4). The problem is then the same as multiplying common fractions. • The product (33/32) is reduced to a mixed number (1 1/32) by dividing 32 into 33. • The final product is 1 1/32.

MULTIPLYING MIXED NUMBERS BY MIXED NUMBERS

The multiplication of a mixed number by a mixed number is performed in the same manner as the multiplication of a mixed number by a fraction except that both mixed numbers are reduced to improper fractions.

Illustration. Multiply 5 3/8" by 2 3/4"

Solution. 5 3/8" x 2 3/4" = 43/8 x 11/4 = 473/32 = 14 25/32"

Explanation. • Reduce both mixed numbers to improper fractions and multiply in the same manner as when multiplying common fractions.

MULTIPLICATION OF MIXED NUMBERS COMBINED WITH OTHER PROCESSES

The process of multiplication of mixed numbers is often combined with other processes in solving problems in carpentry. For example, a combination of processes is used to determine the amount of material needed for *cross bridging*, a method of strengthening floors and distributing the load imposed on them. Cross bridging consists of nailing diagonal pieces between joists, figure 12-2.

Illustration. The pieces used for cross bridging measure 1" x 3" and are 27 1/2" long. How many feet are required to place 2 pieces in each of 10 spaces?

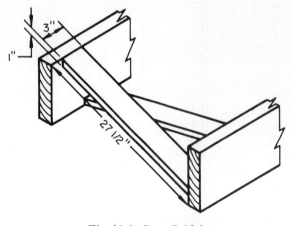

Fig. 12-2 Cross Bridging

Solution.   10 x 2 x 27 1/2″ = 20/1 x 55/2″ = 1100/2 = 550″

$$\begin{array}{r} 45'\ 10'' \\ 12\ \overline{)\ 550} \\ 48 \\ \hline 70 \\ 60 \\ \hline \dfrac{10}{12} = 10'' \end{array}$$

Explanation. • The process involves multiplication of whole and mixed numbers and division of whole numbers. (The measurement of 1″ x 3″ has nothing to do with determining the length of the pieces.) The number of pieces to be used in each space is multiplied by the number of spaces (2 x 10 = 20). • Then 20 is multiplied by the mixed number (27 1/2″). In the process, 27 1/2 is reduced to an improper fraction (55/2). • The multiplication is carried out in the same manner as in the multiplication of common fractions. • To reduce 550″ to feet, 550″ is divided by 12″. The length of the piece needed is 45′ 10″.

Illustration. A dovetail template layout requires 3/4″ of material and a 3/8″ for waste allowance for each dovetail. What is the length of the material required to make 14 dovetail templates?

Solution.   (3/4″ + 3/8″) x 14 = 3/4 + 3/8 L.C.D. = 8

$$\begin{array}{r} 3/4 = 6/8 \\ \underline{3/8 = 3/8} \\ 9/8 \end{array}$$

9/8 x 14/1 = 126/8 = 15 6/8 or 15 3/4″

Explanation. • In solving problems of this kind, first determine the length required to make one dovetail. • Then multiply the number needed by the sum (3/4 + 3/8 = 9/8). • Material required for one template is 3/4″ and 3/8″; therefore, add 3/4″ and 3/8″ to find the total length for one template. • Since it takes 9/8″ for one space, multiply 9/8″ by 14. This product gives the total length of the material needed.

PRACTICAL APPLICATION

In construction, the number of bricks needed to cover a wall is determined by multiplication.

To determine the number of bricks for a wall, multiply each square foot of wall surface by the average number of bricks per square foot: 7 1/2 for a wall 4″ thick, 15 for a wall 8″ thick, and 23 for a wall 12″ thick. For openings that are more than 2′ square, deduct the area in square feet.

Illustration. Determine the number of bricks required to cover the surface of the 4″ wall in figure 12-3, page 68. The wall is 12′ by 14′ with a window 3′ by 3′.

Solution.   12′ x 14′ = 168 sq. ft.   3′ x 3′ = 9 sq. ft.   168 – 9 = 159 sq. ft.

159 sq. ft. x 7 1/2 = 159 x 15/2 = 2385/2 = 1192 1/2 = 1193

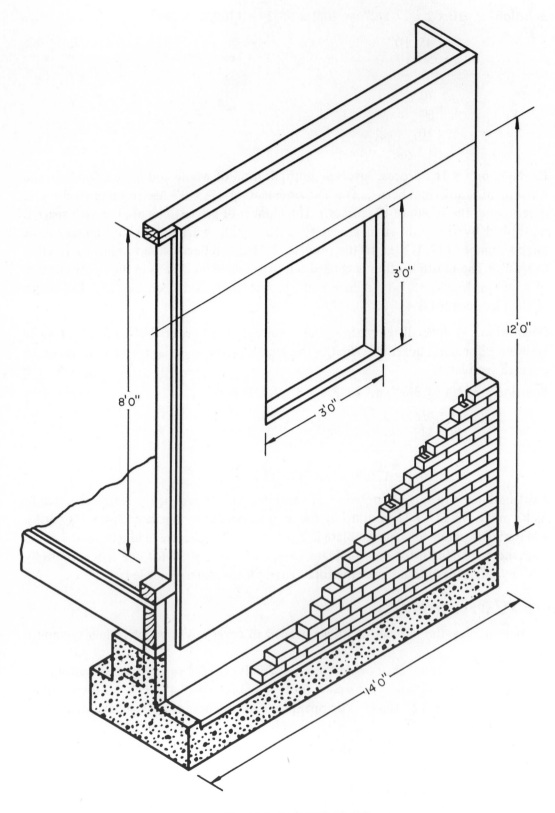

Fig. 12-3 Brick Veneer Wall

Explanation. • To determine the number of bricks needed to cover the wall, first multiply the dimensions to find the total surface area (12' x 14' = 168 sq. ft.). • Deduct the window opening (3' x 3' = 9 sq. ft.). • Since it takes 7 1/2 bricks per square foot of surface wall, multiply 159 by 7 1/2. The result is 1192 1/2 which, when carried to the next whole number, becomes 1193 bricks. • In this problem, the amount of waste from the brick is not considered.

• The size of the floor joists is determined by multiplication. To determine the joist size, first consider the load which is to be placed on each joist. • Read the drawing for the joist span and the distance spaced on center. • Multiply the joist space by the load per square inch.

Illustration. A *floor joist* is a structural member used in a structure to support the floor and loads which are imposed on it. Floor joists are spaced 16" O.C. over a span of 12' in a certain structure. If 50 lb. per square foot is the average load for residential housing, what size joist should be used?

Solution.   16" = 1 1/3' = 4/3'
              4/3 x 12 x 50 = 800 lb.

Explanation. • Change 16" to feet (1 1/3') and then change 1 1/3 to an improper fraction (4/3). • Multiply 4/3 x 12 x 50 which gives 800 lb. Refer to figure 12-4 and locate the figure 12 in the top column, which represents the joist span. • Move down that column to the nearest figure over 800 lb. (980 lb.). • Follow this line to the left to the first column, solid dressed sizes. The figures are 2 x 8, the size of the joist.

Floor Joists (Allowable Fiber Stress 1,200 Lb. Per Sq. In.)

Solid Dressed Sizes (In.)	4-6	7	8	9	10	11	12	13	14	15	16	17	18	19
2 x 6	1,127	961	837	738	660	595	541	494	454	419	388	360	335	312
2 x 8	1,605	1,605	1,503	1,331	1,191	1,075	980	898	827	766	712	662	619	580
2 x 10	2,020	2,020	2,020	2,020	1,912	1,730	1,578	1,449	1,336	1,238	1,153	1,077	1,009	946
2 x 12	2,435	2,435	2,435	2,435	2,435	2,435	2,328	2,136	1,973	1,832	1,708	1,597	1,497	1,410
2 x 12													2,076	1,954

(SPAN IN FEET across top columns)

Fig. 12-4 Joist Sizes Based on Span and Load

APPLICATION

1. Multiply the following mixed numbers by whole numbers as indicated.

   a. 2 1/8 x 12   g. 2 5/6 x 6   m. 5 7/8 x 8
   b. 15 x 2 7/10   h. 15 x 2 3/8   n. 18 x 1 1/2
   c. 5 x 3 9/32   i. 10 1/7 x 64   o. 5 x 2 4/5
   d. 3 1/4 x 16   j. 1 2/3 x 12   p. 6 7/16 x 21
   e. 8 x 3 9/16   k. 14 x 5 1/8   q. 11 x 3 3/5
   f. 6 3/32 x 22   l. 1 5/8 x 3   r. 4 x 4 1/6

2. How many feet of 3/4″ x 10″ material are needed for 12 shelves which are each 5 2/3′ long?

3. What width hole is needed in which to place 3 switch boxes which are each 2 1/4″ wide?

4. Multiply each of the following pairs of numbers.

   a. 1 1/2 x 1/2   h. 3/10 x 5/6   o. 4 1/8 x 2 13/16
   b. 1/3 x 7 1/2   i. 2 1/2 x 9/16   p. 2 2/3 x 3 3/8
   c. 11/32 x 15 1/2   j. 5/6 x 12 2/3   q. 1 3/4 x 3 1/7
   d. 32 1/2 x 1/64   k. 5/8 x 10 11/12   r. 3 2/5 x 4 3/7
   e. 2 1/4 x 5/6   l. 3/7 x 2 1/2   s. 2 1/16 x 3 3/4
   f. 3/4 x 4 1/2   m. 2 3/4 x 4 1/2   t. 3 3/4 x 2 1/5
   g. 9/32 x 9 3/8   n. 2 2/3 x 5 5/8

5. How many bricks are needed for a foundation wall which measures 12 1/2″ x 3 3/4′ x 22 2/3′?

6. When making the center tap joint shown below, remove 1 1/4″ of insulation from the main wire and 1 1/2″ from the wire which is to be attached to the main wire. If 25 connections are to be made, how many inches of insulation are removed?

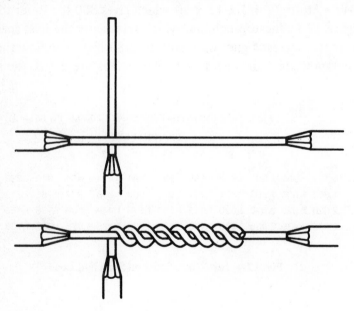

Unit 13
Division of Mixed Numbers

OBJECTIVES

After studying this unit, the student should be able to

- divide mixed numbers by whole numbers, whole numbers by mixed numbers, and mixed numbers by mixed numbers.
- divide mixed numbers using parentheses.

When a mixed number is divided by a whole number, fraction, or another mixed number, the mixed numbers are reduced to improper fractions, the terms of the divisors are inverted, and multiplication is used.

DIVIDING A WHOLE NUMBER BY A MIXED NUMBER

To determine how many times a mixed number is contained in another number, use the process of division.

Illustration. A carpenter needs to know how many pieces of lumber 2 1/2′ long can be cut from a piece 15′ long.

Solution. Divide 15′ by 2 1/2′.

$$15 \div 2\,1/2 = 15/1 \div 5/2 = \frac{\overset{3}{15}}{1} \times \frac{2}{\underset{1}{5}} = \frac{3}{1} \times \frac{2}{1} = \frac{6}{1} = 6 \text{ pieces}$$

Explanation. • When 2 1/2 is changed to an improper fraction, it becomes 5/2. The denominator of the whole number is 1. • The divisor (5/2) is inverted and becomes 2/5. • Multiplying 15/1 by 2/5 gives 6 pieces.

DIVIDING A MIXED NUMBER BY A WHOLE NUMBER

Division of a mixed number by a whole number is often used to achieve equal spacing of studs, figure 13-1.

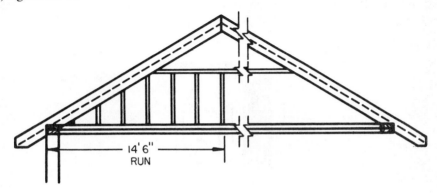

Fig. 13-1 Gable End Studs

Illustration. A concrete walk with a total length of 14′ 7″ is to be constructed from a house to a driveway. The driveway is to be poured in 7 sections of equal length. What is the length of each section?

Solution. 14′ 7″ ÷ 7 = 14 7/12′ ÷ 7/1 = 175/12′ x 1/7 = 175/84 = 2 1/12′ long

Explanation. A mixed number is divided by a whole number. The denominator of the whole number is 1. Reduce 14′ 7″ to an improper fraction (175/12) and invert the divisor (7/1 to 1/7). Multiply 175 x 1 to obtain a new numerator and 12 x 7 to obtain a new denominator. The product is 175/84. Reduce 175/84 to a mixed number; the final result is 2 1/12′ (2′ 1″).

DIVIDING A MIXED NUMBER BY A FRACTION

To divide a mixed number by a fraction, reduce the mixed number to an improper fraction and invert the terms of the divisor. After this has been done, the process becomes multiplication. After multiplying, reduce the product to a whole or mixed number.

Illustration. A box of 3/4″ brads weighs 1/2 lb. How many boxes are there in 21 1/2 lb.?

Solution. 21 1/2 ÷ 1/2 = 43/2 x 2/1 = 86/2 = 43

Explanation. • Read the problem carefully. Notice that 3/4″, the size of the brads, has nothing to do with solving the problem. • The first step is to reduce 21 1/2 to an improper fraction (43/2). • Invert the divisor (1/2 to 2/1). • Multiply the numerator by the numerator and the denominator by the denominator. • The result is 86/2 or 43. This illustrates that there are 43 boxes of brads in 21 1/2 lb., with each box weighing 1/2 lb.

DIVIDING A MIXED NUMBER BY A MIXED NUMBER

When a mixed number is to be divided by a mixed number, reduce both mixed numbers to improper fractions and invert the divisor. The process is then multiplication.

Illustration. A low-voltage transformer (a device used to increase or decrease voltage) may be used to operate remote-control relays in residences, figure 13-2. Assume that a 24-volt transformer weighs 2 1/4 lb. How many transformers will weigh 13 1/2 lb.?

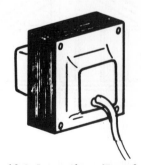

Fig. 13-2 Low-voltage Transformer

Solution. $13\ 1/2 \div 2\ 1/4 = 27/2 \div 9/4$

$$\frac{\overset{3}{\cancel{27}}}{\underset{1}{\cancel{2}}} \times \frac{\overset{2}{\cancel{4}}}{\underset{1}{\cancel{9}}} \quad \frac{6}{1} = 6$$

Explanation. • When 13 1/2 and 2 1/4 are reduced to improper fractions, the results are 27/2 and 9/4. • Invert the terms of the divisor to obtain 4/9. • After cancellation, the problem is 3/1 x 2/1. The final result is 6.

USING PARENTHESES IN DIVISION OF MIXED NUMBERS

Parentheses are used in the division of mixed numbers to indicate a grouping of numbers. The parentheses indicate that the enclosed numbers are considered a single number. The result of the numbers inside the parentheses should be determined before the rest of the problem is solved.

Illustration. $1\ 7/8 \div (3\ 3/4 \div 2\ 2/5)$

Solution. $1\ 7/8 \div (3\ 3/4 \div 2\ 2/5) = 15/8 \div (15/4 \div 12/5) = 15/8 \div (15/4 \times 5/12) =$

$$\frac{15}{8} \div \left(\frac{\overset{5}{\cancel{15}}}{4} \times \frac{5}{\underset{4}{\cancel{12}}} \right) = \frac{15}{8} \div \frac{25}{16} = \frac{\overset{3}{\cancel{15}}}{\underset{1}{\cancel{8}}} \times \frac{\overset{2}{\cancel{16}}}{\underset{5}{\cancel{25}}} = \frac{3 \times 2}{1 \times 5} = \frac{6}{5}$$

$$6/5 = 1\ 1/5$$

Explanation. • The dividend (1 7/8) is reduced to an improper fraction (15/8). To change the numbers within the parentheses to a single number, 3 3/4 and 2 2/5 are reduced to improper fractions (15/4 and 12/5). • In the next step, the terms of the enclosed divisor are inverted (12/5 to 5/12) and the sign is changed from division to multiplication (15/4 x 5/12). • The fractions are reduced to lowest terms by cancellation (5/4 x 5/4). • The result (25/16) is the regular divisor. • The problem now reads 15/8 ÷ 25/16. • Invert the divisor (25/16) and change the sign to multiplication. The result is 1 1/5.

APPLICATION

1. Divide the following whole numbers by mixed numbers.

a. $4 \div 1\ 1/2$	i. $7 \div 2\ 2/3$	q. $18 \div 2\ 2/3$
b. $68 \div 4\ 2/5$	j. $14 \div 2\ 5/16$	r. $56 \div 18\ 3/4$
c. $27 \div 4\ 1/4$	k. $69 \div 3\ 4/5$	s. $64 \div 3\ 3/64$
d. $19 \div 5\ 1/4$	l. $33 \div 2\ 1/8$	t. $120 \div 12\ 1/2$
e. $8 \div 2\ 1/4$	m. $5 \div 2\ 2/5$	u. $12 \div 5\ 1/2$
f. $40 \div 3\ 1/3$	n. $10 \div 2\ 7/8$	v. $8 \div 2\ 3/4$
g. $57 \div 3\ 3/8$	o. $39 \div 3\ 3/8$	w. $27 \div 1\ 1/4$
h. $20 \div 6\ 1/2$	p. $11 \div 3\ 4/5$	x. $134 \div 3\ 11/64$

2. A bundle of metal weighs 156 lb. One sheet of the metal weighs 1 1/2 lb. How many sheets are there in the bundle?

3. A roll of 3/8″ copper tubing is 25 1/2′ long. How many pieces of tubing 8 1/2′ long can be cut from the roll?

4. A *sole plate* is a horizontal structural member which supports the bottom end of studs. The sole plate in the figure below is 48′ long with studs placed 1 1/3′ O.C. How many equal spaces are there in the 48′?

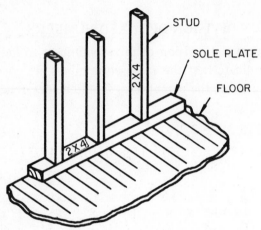

5. A carpenter works a total of 47 2/3 hours in 6 1/2 days. How many hours did he work on the average per day?

6. If the average time for installing a lock on an outside door is 2 1/3 hours, how many locks can be installed in 8 hours?

7. Divide the following mixed numbers.

a. 15 3/8 ÷ 3 7/8   h. 5 2/3 ÷ 2 3/5   o. 51 1/4 ÷ 5 3/16
b. 4 1/5 ÷ 2 1/3   i. 2 13/16 ÷ 3 1/4   p. 3 1/16 ÷ 2 3/8
c. 3 1/4 ÷ 4 3/8   j. 4 3/8 ÷ 4 1/5   q. 2 17/32 ÷ 3 5/8
d. 8 1/2 ÷ 2 1/4   k. 3 2/3 ÷ 3 2/3   r. 5 9/10 ÷ 3 3/5
e. 2 5/8 ÷ 2 7/32   l. 19 1/3 ÷ 2 5/6   s. 7 13/16 ÷ 7 13/16
f. 2 3/5 ÷ 4 1/5   m. 12 1/4 ÷ 3 1/2   t. 2 2/3 ÷ 2 7/16
g. 15 1/2 ÷ 4 5/8   n. 3 1/12 ÷ 4 3/4   u. 4 3/8 ÷ 2 5/6

8. A carpenter removes 2 3/4″ of insulation from the ends of both wires when making a *western union splice*, shown in the figure below. A total of 137 1/2″ of insulation is removed. How many western union splices does he make?

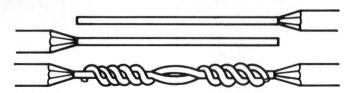

9. When making a *pigtail splice,* illustrated below, 1 1/2" of insulation are removed from the ends of the wires being spliced. A total of 31 1/4" of insulation is removed (1 1/2" from each wire). How many wires are there?

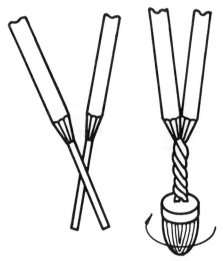

10. There are 90 duplex receptacles, shown below, in 7 1/2 boxes. Find the average number of receptacles in each box.

11. How many pieces of 1/4" tubing 8 3/4" long can be cut from a piece which is 50 1/4" long?

12. A total of 7 1/2 standard-sized bricks are required to cover 1 sq. ft. on a wall 4" thick. What is the measurement of a wall (in square feet) which requires 1200 bricks?

13. A carpenter finishes 190 square feet of concrete in 4 3/4 hours. What is the average number finished per hour?

14. If a board 16 1/4′ long is cut into 4 pieces, what is the length of each piece?

15. What are the actual dimensions of a room which is drawn to a scale of 1/4″ = 1′ 0″, and measures 2 3/4″ by 3 1/4″?

16. The scale on the blueprints for a metal shop is 1/8″ = 1′ 0″, shown below. Find the actual dimensions of the walls of the shop.

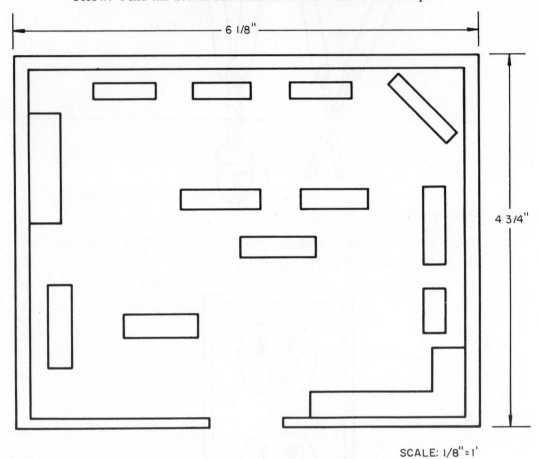

SCALE: 1/8″=1′

17. How many pieces of reinforcing rods 3 1/2′ long can be cut from 40 metal rods which are each 21′ long?

18. Divide the following numbers as indicated.

a. (1 7/8 ÷ 1 3/16) ÷ 1 21/32
b. (2 3/4 ÷ 3 4/5) ÷ 2 3/8
c. (6 1/3 ÷ 3 1/5) ÷ 3 3/8
d. (3 7/8 ÷ 4 1/2) ÷ 3 3/5
e. (3 3/4 ÷ 1 2/3) ÷ 4/5
f. (7 1/3 ÷ 4 2/3) ÷ 3 3/8
g. 4 3/4 ÷ (3 3/4 x 4 1/4)
h. (3 3/32 ÷ 2/3) ÷ 3/64

i. 2 9/10 ÷ (2 3/5 ÷ 3 3/4)
j. 4 3/8 ÷ (2 5/6 ÷ 3 7/12)
k. 5 1/2 ÷ (11/16 ÷ 3 3/4)
l. 3/5 x (4 1/2 ÷ 2 3/4)
m. 2/5 x (4 1/2 ÷ 2 3/4)
n. 3/4 ÷ (2/3 x 4 1/3)
o. (6 3/16 + 3 7/8) ÷ 2 1/2
p. 5/8 ÷ (1/2 x 3 1/2)

Unit 14   Conversion of Decimal Fractions to Common Fractions

OBJECTIVES

After studying this unit, the student should be able to

- read and write decimals.
- read a micrometer in making measurements.
- change decimal fractions to common fractions.

UNDERSTANDING THE DECIMAL POINT

Dimensions of structures are sometimes given as decimal fractions. Any fraction having a multiple of 10 for the denominator, such as 10, 100, or 1000, is called a *decimal fraction*. It is written without the denominator by using a *decimal point,* (.) placed in front of the numerator. The decimal point separates a whole number from a fraction.

The fractional part of a number is on the right of the decimal point and the whole number is on the left of the decimal point. The number of figures on the right of the decimal point indicates the number of zeros in the denominator; for example, the denominator of the decimal fraction .5 contains one zero (10). The decimal fraction .5 is equal to the common fraction 5/10. The value of a decimal does not change when zeros are added after it; 1.5 has the same value as 1.50

There are two kinds of decimal fractions used in construction problems, pure decimals and mixed decimals. When there is no whole number in the fraction, as in .5 or .50, the number is called a *pure decimal*. A *mixed decimal* is the combination of a whole number and a decimal, as in 1.5 or 2.512.

READING DECIMAL FRACTIONS

The first step in reading decimals is to learn the names of the decimal places, figure 14-1.

```
.5        =      5/10    = 5 tenths
.05       =      5/100   = 5 hundredths
.005      =      5/1000  = 5 thousandths
.0005     =    5/10000   = 5 ten-thousandths
.00005    =   5/100000   = 5 hundred-thousandths
.000005   = 5/1000000    = 5 millionths
```

Fig. 14-1 Reading Decimal Fractions

Read the number as though it were a whole number. Begin at the number (or zero) to the right of the decimal point and express the proper unit place, such as tenths, hundredths, or thousandths. The last figure to the right indicates the decimal place of the number.

Illustration. Read the following decimal fractions: .5, .18, .026, and 22.0556.

Solution. .5, five tenths; .18, eighteen hundredths; .026, twenty-six thousandths; 22.0556, twenty-two and five hundred fifty-six ten-thousandths.

Explanation. • In .5, the 5 occupies the tenths place so it is read as five tenths. • The last right-hand figure (8) occupies the place of hundredths, so .18 reads as eighteen hundredths. • The last figure to the right of the decimal fraction .026 occupies the place of the thousandths; it is read as twenty-six thousandths. • The whole number (22) is read. The word *and* is correctly used, but the word *point* is often used to indicate the decimal point. The last figure in the decimal part of the number is 6, which occupies the ten-thousandths place. The decimal part of the number is read as five hundred fifty-six ten-thousandths. The complete number (22.0556) is read twenty-two and five hundred, fifty-six ten-thousandths.

WRITING DECIMAL FRACTIONS

When writing a decimal fraction, write the number as though it were a whole number and place the decimal point so the last figure at the right is the place named by the decimal fraction.

Illustration. Write the following as decimal fractions: eight tenths, fifteen hundredths, one hundred twenty-one thousandths, fifteen and two hundred thirty-three ten-thousandths.

Solution. .8, .15, .121, 15.0233.

Explanation. • The first figure at the right of the decimal point is the tenths place, .8. • With two numbers right of the decimal, the figure indicates hundredths, .15. • Figures having three numbers right of the decimal are called thousandths, .121. • The figures to the left of the decimal point are whole numbers and are written as whole numbers, followed by a decimal point. The decimal fraction is then written. The last figure in the decimal is 3 which occupies the ten-thousandths place. The number is written as 15.0233.

PRACTICAL APPLICATION: READING THE MICROMETER

The *micrometer* is a precision tool used to measure minute, or very small, distances. It is used in carpentry to measure within a thousandth of an inch. The outside micrometer in figure 14-2 is used to measure the outside diameters or thickness of an object. (The *diameter* of a circle or round object is the distance straight across the center.) Each standard outside micrometer has a size range of 1″. It is possible to obtain a set of five micrometers covering sizes 0-1″, 1-2″, 2-3″, 3-4″, and 4-5″.

An inside micrometer, figure 14-3, is used to measure inside distances. Inside and outside micrometers are read in the same manner.

The spindle of the micrometer has 40 threads per inch and moves 1/40 of an inch, or .025″, per revolution. Each revolution (.025) is equal to one of the divisions on the sleeve. Four revolutions equal .100, indicated by the *1* just above the index line. Every four divisions on the sleeve are marked 1, 2, 3, 4, and so on. These markings represent 0.100″,

0.200″, 0.300″, 0.400″, and so on. The beveled edge of the thimble is graduated into 25 equal parts with each fifth numbered 0, 5, 10, 15, or 20. This assures easy reading of detailed measurements. When 25 of these graduations pass the index line, the spindle has traveled .025″. If the thimble moves one graduation, it has moved .001″.

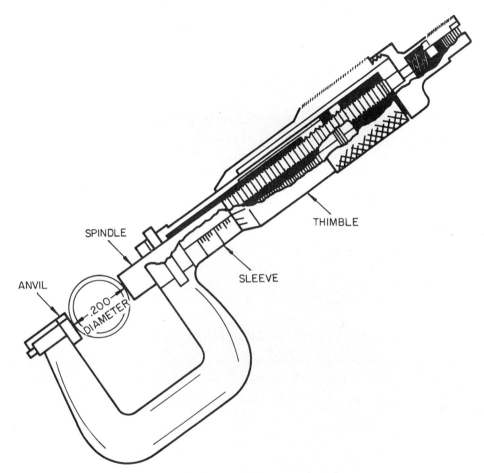

Fig. 14-2 Outside Micrometer

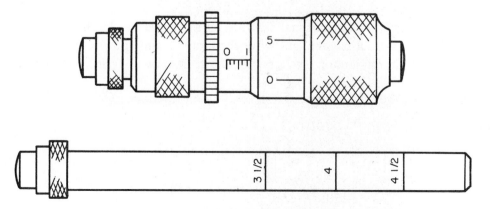

Fig. 14-3 Inside Micrometer

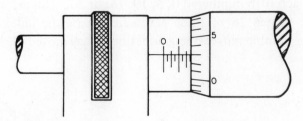

Fig. 14-4 Reading of 0.178″ on a 0-1″ Micrometer

To read the micrometer, first note the last or highest number visible on the sleeve. In figure 14-4 the largest visible number is 1, which represents 0.100″. Next count the number of 0.025″ divisions visible between the *1* and the thimble edge. In this case, three of the 0.025″ divisions are visible beyond the *1*. This represents 0.075″. Now note the number on the thimble that lines up with the horizontal reference line on the sleeve. The number 3, which represents 0.003″, is directly across from the horizontal reference line in this example. Finally, add the three numbers expressed in thousandths (0.100″, 0.075″, and 0.003″) to find the exact size of the work. Since the micrometer illustrated measures from 0 to 1″, the size of the work is 0.178″. If this reading were obtained on a 3-4″ micrometer, the size would be 3.178″.

CHANGING DECIMALS TO COMMON FRACTIONS

To change a decimal to a common fraction, use the figure or figures in the decimal as the numerator and write 1 in the denominator. Then add as many zeros to the denominator as there are figures in the numerator. Omit the decimal point and reduce to lowest terms.

Illustration. Change the decimal .6 to a common fraction.

Solution. .6 = 6/? = 6/10

$$\frac{\overset{3}{\cancel{6}}}{\underset{5}{\cancel{10}}} = \frac{3}{5}$$

Explanation. • Write the decimal figure 6 as a numerator, leaving off the decimal point. • Draw a fraction line and write the figure 1 and add one zero since there is one figure in the numerator. • Reduce 6/10 to its lowest terms. The result is 3/5.

Illustration. Change the decimal .026 to a common fraction.

Solution. .026 = 026/? = 026/1000

$$\frac{\overset{13}{\cancel{26}}}{\underset{500}{\cancel{1000}}} = \frac{13}{500}$$

Explanation. • Place 026 as the numerator and 1000 for the denominator since there are three numbers in the numerator. • The fraction is reduced to 13/500.

It is sometimes necessary to change a decimal fraction of a foot into a fraction of an inch.

Illustration. Change 0.375′ to inches.

Solution. 0.375 ft. = 375′/1000 = 3/8′

$$\begin{array}{r} .375' \\ \times\ \ 12'' \\ \hline 750 \\ 375\ \ \ \ \\ \hline 4.500 = 4\ 1/2'' \end{array}$$
   or   3/8′ x 12″ = 36/8 = 4 1/2″

Explanation. • There are two methods that may be used to solve this problem. • In the first solution, .375′ is multiplied by 12″. • The result is 4.5″ or 4 1/2″. • In the second solution, .375′ is changed to a common fraction, 3/8′ and the 3/8′ is multiplied by 12″. • The result is 4 1/2″.

CHANGING MIXED DECIMALS TO MIXED NUMBERS

Two procedures may be used to change a mixed decimal to a mixed number.

Procedure 1: Write a fraction which uses the whole number and the decimal fraction figures as the numerator; and a multiple of ten, such as 10, 100, and 1000, which corresponds to the place value of the last figure of the decimal as the denominator. Reduce the fraction to its lowest terms and change it to a mixed number.

Illustration. Change the decimal 2.210 to a mixed number.

Solution. Step 1. 2.210 = 2210/? = 2210/1000

   Step 2. 2210/1000 = 221/100 = 2 21/100

Explanation. • The numerator of the fraction is 2210 and the denominator is 1000 since there are three zeros to the right of the decimal point. • The improper fraction (2210/1000) is reduced and then expressed as a mixed number.

Procedure 2: Write the figures following the decimal point as the numerator of the fraction. Use the multiple of ten that corresponds to the place value as the denominator. Reduce the fraction to its lowest terms and annex it to the whole number.

Illustration. Change 5.25 to a mixed number.

Solution. Step 1. .25 = 25/100

   Step 2. 25/100 = 1/4

   Step 3. 5.25 = 5 1/4

Explanation. • In this case, .25 requires 100 as the denominator (25/100). • When it is reduced to its lowest terms (1/4), it is annexed to the whole number (5). • The answer is 5 1/4.

CHANGING DECIMALS TO COMMON FRACTIONS WITH GIVEN DENOMINATORS

Sometimes it is necessary to change a decimal to a common fraction which has a given denominator.

To change a decimal to a fraction which has a given denominator, multiply the decimal by the fractional equivalent of 1, expressed in terms of the required denominator.

Illustration. Change 0.25 to a common fraction with 4 as the denominator.

Solution. 0.25 x 4/4 = 1.00/4 = 1/4

Explanation. • Multiply the decimal (0.25) by the fractional equivalent of 1 (4/4). (The fraction 4/4 expresses 1 in terms of the required denominator (4).) • The answer is 1/4.

APPLICATION

1. Express the following decimals in words.

.1	.6	.02	.07	.003	.008	.37
.2	.7	.03	.08	.004	.009	.85

2. Express the following decimals in numbers.
   a. two-tenths
   b. three-tenths
   c. one-tenth
   d. seven-hundredths
   e. twenty-five hundredths
   f. nineteen ten-thousandths
   g. two and eighteen ten-thousandths
   h. six hundred seventy-five and twenty-five hundredths
   i. six-millionths
   j. nine-tenths
   k. eight-thousandths
   l. two hundred and fifty-eight thousandths

3. Change the following decimals to common fractions and reduce to lowest terms.

a. .25	d. .19	f. .345	h. .750	j. .0045	l. .0150	n. .3750
b. .50	e. .225	g. .075	i. .900	k. .0875	m. .0946	o. .5575
c. .45						

4. Change the following mixed decimals to mixed fractions.

a. 2.375	c. 3.964	e. 5.267	g. 1.1562	i. 5.0054	k. 5.00084
b. 2.248	d. 2.132	f. 3.125	h. 2.0025	j. 2.20625	l. 4.35625

5. Change each of the following decimals to common fractions. Use the given denominators.

   a. .875 to 64ths   b. .875 to 8ths   c. .125 to 8ths

Unit 15   Conversion of
Common Fractions to Decimal Fractions

OBJECTIVES

After studying this unit, the student should be able to

- change common fractions to decimal fractions.
- change mixed numbers to mixed decimals.
- round off decimal fractions.
- change inches and fractions of an inch to feet and decimal fractions of a foot.

One of the processes for changing a fraction to a decimal fraction is to divide the numerator of the fraction by its denominator.

Since in division the dividend must have more digits than the divisor, the division of proper fractions calls for the addition of one or more zeros to the numerator.

Illustration. Change the common fraction 1/2 to a decimal fraction.

Solution.
$$1/2 \;=\; 2\overline{)1.0}^{\,.5}\;$$
$$\underline{1\,0}$$

Explanation. • The numerator (1), which is used as the dividend, is smaller than the denominator (2). • Therefore, a decimal point and one zero are added. • The decimal point in the quotient is placed exactly above the point in the dividend. • The division is then like that of whole number division. • The result is .5.

The procedure used when two zeros must be added to the dividend is similar to the procedure just discussed.

Illustration. Change 3/4 to a decimal fraction.

Solution.
$$3/4 \;=\; 4\overline{)\,3.00}^{\,.75}$$
$$\underline{2\,8}$$
$$20$$
$$\underline{20}$$

Explanation. • The procedure is the same as in the first illustration except that two zeros are annexed instead of one. • The fraction 3/4 is equivalent to .75. • There is no remainder.

Some fractional problems have answers that constantly repeat themselves, regardless of how many zeros are added to the dividend.

Illustration. Change 2/3 to a decimal fraction.

Solution.   .6666 = .66 2/3

$$3\overline{)2.0000}$$

    1 8
    ――
     20
     18
    ――
     20
     18
    ――
     20
     18
    ――
      2

Explanation. • Division of the fraction 2/3 is carried out to four places. • The process is simply repeated if it is carried any further.

CHANGING MIXED NUMBERS TO MIXED DECIMALS

To change a mixed number to a mixed decimal, write the numerator of the fractional part of the number; place a decimal point after it; annex zeros and divide by the denominator. After this is done, annex the answer to the whole number.

Illustration. Change 5 1/4" to a mixed decimal.

Solution.   5 1/4 = 5.25   .25

$$4\overline{)1.00}$$

                 8
                ――
                20
                20

Explanation. • The decimal value of 1/4 is found by dividing 1 by 4. • The decimal which results, .25, is annexed to the whole number 5. • The number is expressed as 5.25.

In order to add measurements as mixed decimals it is necessary to change the fractions to decimals. As has been illustrated, to change fractions to decimals, the numerator is divided by the denominator.

Illustration. Add the following measurements as mixed decimals: 5 3/4", 10 1/5", 14 1/8", 3 7/8", 2 3/8", and 9 3/5".

Solution.   5 3/4" =   5.750"
                 10 1/5" =   10.200"
                 14 1/8" =   14.125"
                  3 7/8" =   3.875"
                  2 3/8" =   2.375"
                  9 3/5" =   9.600"
                            ――――――――
                             45.925"

Explanation. • The decimal value of each fractional part of each whole number is found by dividing the numerator by the denominator. • The division is carried out to three places.

FRACTION-DECIMAL EQUIVALENTS

To be familiar with fraction-decimal equivalents, study the following chart which gives decimal equivalents for the fractional parts of an inch.

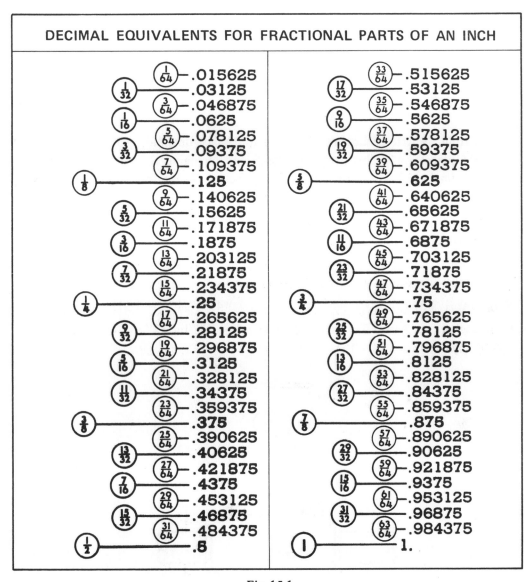

DECIMAL EQUIVALENTS FOR FRACTIONAL PARTS OF AN INCH

$\frac{1}{64}$ — .015625		$\frac{33}{64}$ — .515625	
$\frac{1}{32}$ — .03125		$\frac{17}{32}$ — .53125	
$\frac{3}{64}$ — .046875		$\frac{35}{64}$ — .546875	
$\frac{1}{16}$ — .0625		$\frac{9}{16}$ — .5625	
$\frac{5}{64}$ — .078125		$\frac{37}{64}$ — .578125	
$\frac{3}{32}$ — .09375		$\frac{19}{32}$ — .59375	
$\frac{7}{64}$ — .109375		$\frac{39}{64}$ — .609375	
$\frac{1}{8}$ — .125		$\frac{5}{8}$ — .625	
$\frac{9}{64}$ — .140625		$\frac{41}{64}$ — .640625	
$\frac{5}{32}$ — .15625		$\frac{21}{32}$ — .65625	
$\frac{11}{64}$ — .171875		$\frac{43}{64}$ — .671875	
$\frac{3}{16}$ — .1875		$\frac{11}{16}$ — .6875	
$\frac{13}{64}$ — .203125		$\frac{45}{64}$ — .703125	
$\frac{7}{32}$ — .21875		$\frac{23}{32}$ — .71875	
$\frac{15}{64}$ — .234375		$\frac{47}{64}$ — .734375	
$\frac{1}{4}$ — .25		$\frac{3}{4}$ — .75	
$\frac{17}{64}$ — .265625		$\frac{49}{64}$ — .765625	
$\frac{9}{32}$ — .28125		$\frac{25}{32}$ — .78125	
$\frac{19}{64}$ — .296875		$\frac{51}{64}$ — .796875	
$\frac{5}{16}$ — .3125		$\frac{13}{16}$ — .8125	
$\frac{21}{64}$ — .328125		$\frac{53}{64}$ — .828125	
$\frac{11}{32}$ — .34375		$\frac{27}{32}$ — .84375	
$\frac{23}{64}$ — .359375		$\frac{55}{64}$ — .859375	
$\frac{3}{8}$ — .375		$\frac{7}{8}$ — .875	
$\frac{25}{64}$ — .390625		$\frac{57}{64}$ — .890625	
$\frac{13}{32}$ — .40625		$\frac{29}{32}$ — .90625	
$\frac{27}{64}$ — .421875		$\frac{59}{64}$ — .921875	
$\frac{7}{16}$ — .4375		$\frac{15}{16}$ — .9375	
$\frac{29}{64}$ — .453125		$\frac{61}{64}$ — .953125	
$\frac{15}{32}$ — .46875		$\frac{31}{32}$ — .96875	
$\frac{31}{64}$ — .484375		$\frac{63}{64}$ — .984375	
$\frac{1}{2}$ — .5		1 — 1.	

Fig. 15-1

ROUNDING OFF DECIMALS

When decimals are rounded off, figures which follow the ones necessary to express the degree of accuracy required by the problem are dropped. The degree of accuracy required determines whether decimals are rounded off to tenths, hundredths, or thousandths.

To round off decimals to the required decimal place, the figure to the right of the last figure which is to be used must be considered. If this figure is 4 or less, it is dropped and does not affect the final figure. If the figure is 5 or more, it is dropped and the final figure is increased by 1.

Illustration. Change 7/16 to its decimal equivalent and round off.

Solution.

$$.4375 = .438 \text{ or } .44$$

$$7/16 = 16\overline{)7.0000}$$
$$\underline{6\,4}$$
$$60$$
$$\underline{48}$$
$$120$$
$$\underline{112}$$
$$80$$
$$\underline{80}$$

Explanation. • After changing 7/16 to a four-place decimal (.4375), it is rounded off to three places (.438) by dropping the 5 and changing the last figure (7) to an 8. • The decimal .438 is rounded off to two places (.44) by dropping the 8 and changing the three to 4. • If a high degree of accuracy is required, the best answer is .4375.

In carpentry problems, it is sometimes necessary to change fractional parts of measurements and weights to decimal fractions.

Illustration. The inside diameter of the globe valve shown in figure 15-2 is 4 3/4″. Express the diameter as a decimal fraction of a foot.

Solution. 4 3/4″ = 4.75   4.75″/12″ = 0.312′

Explanation. • Change 3/4″ to .75″. • Divide 4.75″ by 12″. • The answer is 0.3125′. (Divide by 12 since 12″ = 1′.)

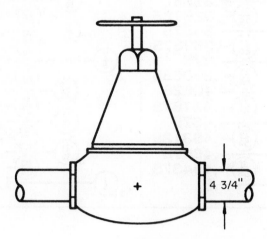

Fig. 15-2 Globe Valve

Illustration. Change 25 1/8″ to a decimal fraction. Then express the figure in feet.

Solution. 25 1/8″ = 25.125   25.125″/12″ = 2.0938′

Explanation. • The second illustration is the same as the first except that in the second illustration, the whole number is larger than 1′. • The 2 in the answer represents 2′ while .0938 is a fraction of a foot.

Illustration. Change 15 oz. to a decimal fraction of a pound.

Solution.   16 oz. = 1 lb.   15 oz. = 15/16 lb.

$$
15/16 = 16\overline{\smash{)}15.0000}\ \ .9375
$$

```
                .9375
15/16 =  16)15.0000
             14 4
               60
               48
              120
              112
               80
               80
```

Explanation. • The procedure is the same as the first three except that 16 is used as the divisor since 16 oz. equals 1 lb. • The answer is .9375 lb.

APPLICATION

1. Change the following mixed numbers to mixed decimals. Carry each answer to four or more decimal places.

a. 1 1/2	j. 5 17/64	s. 5 1/64	ab. 9 9/1280
b. 5 5/16	k. 7 3/16	t. 5 11/1280	ac. 7 7/160
c. 11 7/32	l. 4 3/32	u. 3 9/160	ad. 7 7/320
d. 5 1/8	m. 4 3/8	v. 9 9/320	ae. 11 7/640
e. 2 15/32	n. 10 1/4	w. 5 9/640	af. 5 9/100
f. 2 9/32	o. 5 11/64	x. 11 5/640	ag. 3 3/80
g. 7 13/64	p. 6 5/64	y. 8 1/20	ah. 2 3/160
h. 5 7/64	q. 2 1/16	z. 10 1/40	ai. 3 13/1280
i. 2 13/32	r. 7 1/32	aa. 6 1/80	aj. 2 11/100

2. Change the following fractions to decimals. Carry each answer to at least four decimal places.

a. 3/4	h. 7/36	o. 11/39	v. 13/19
b. 2/25	i. 7/8	p. 2/3	w. 11/32
c. 1/64	j. 1/8	q. 5/16	x. 2/9
d. 11/36	k. 11/48	r. 13/16	y. 5/6
e. 5/8	l. 9/15	s. 13/32	z. 7/11
f. 3/8	m. 15/32	t. 5/9	aa. 11/17
g. 7/48	n. 3/64	u. 11/16	ab. 4/19

3. A hole in the piece of metal in the figure below is 3/8'' in diameter. What is its size expressed as a decimal?

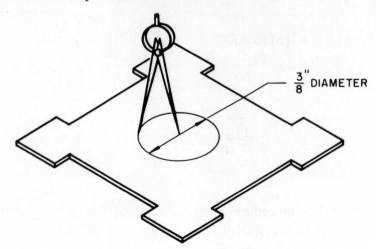

$\frac{3}{8}''$ DIAMETER

4. A hole in a piece of metal is 1 1/4'' in diameter. What is its size expressed in decimal form?

5. The joint of the plastic pipe in the figure is 1 3/8'' in outside diameter. Express the measurement in decimal form.

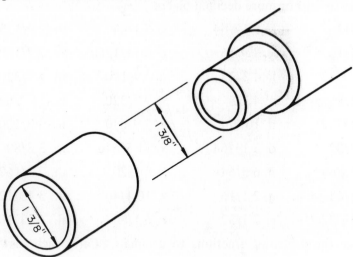

6. Which is greater, .583'' or 17/32''?

7. Which is greater, .453'' or 31/64''?

8. Change 3 1/2'' to a decimal fraction of a foot.

9. Change 9 3/4'' to a decimal fraction of a foot.

10. Change 26 1/2'' to feet and a decimal fraction of a foot.

11. Change 30 1/8'' to feet and a decimal fraction of a foot.

12. The measurement of metal with a micrometer is illustrated in the figure below. The thickness of the piece is 1/8″. What is the measurement expressed in decimal form which would be read on the micrometer?

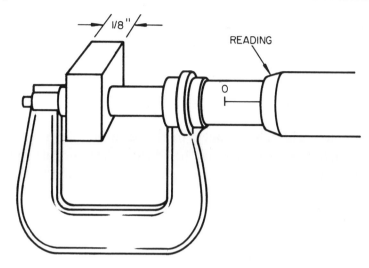

13. Make a table of at least 25 basic fractions and their decimal equivalents.

14. Change 9 oz. to a decimal fraction of a pound.

15. Change 12 oz. to a decimal fraction of a pound.

16. Change 3 lb. 13 oz. to a decimal fraction of a pound.

Unit 16
Addition of Decimals

OBJECTIVES

After studying this unit, the student should be able to

- add pure and mixed decimals.
- add combinations of common fractions and decimal fractions.

The addition of decimals is very similar to the addition of whole numbers. It was noted in Unit 2 that only like quantities may be added. The same rule applies to the addition of decimals. For example, tenths must be added to tenths to give tenths.

ADDING PURE DECIMALS

Pure decimals are decimals without whole numbers. For example, .5 is a pure decimal; 3.5 is classified as a mixed decimal. To add pure decimals, write the numbers in a vertical column so that the decimal points are in line with each other. This places tenths under tenths, hundredths under hundredths, and so on. One column at a time is added, as in the addition of whole numbers. After all the columns have been added, place the decimal point of the sum directly below the decimal point of the addends.

Illustration. Find the sum of .4 and .5.

Solution.   .4
              + .5
              ‾‾‾‾
                .9

Explanation. • The .4 and .5 are pure decimals since there are no figures to the left of the decimal point. • Place the numbers in a vertical column with one decimal point directly under the other decimal point. • After adding .4 and .5, place the decimal point of the sum directly below the decimal points of the addends. • Sometimes the sum of two or more pure decimals is a mixed decimal: $.9 + .5 = 1.4$.

ADDING MIXED DECIMALS

To add mixed decimals, write the numbers in a vertical column so that the decimal points are in line. Then add as in the addition of whole numbers. Place the decimal point of the sum directly below the other decimal points.

Illustration. What is the total length of the metal in figure 16-1?

Solution.   1.5″
               2.38″
            + 4.5881″
            ‾‾‾‾‾‾‾‾‾
              8.4681″

Explanation. • Add as in the addition of whole numbers and place the decimal point of the sum directly below the decimal point of the addends. • Note that the number 8

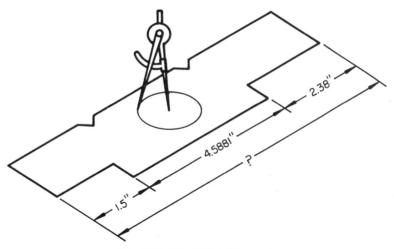

Fig. 16-1 Sheet Metal Layout

at the left of the decimal point is a whole number and that the number .4681 on the right is a decimal fraction.

ADDING WHOLE NUMBERS AND DECIMALS

The addition of whole numbers and decimals is similar to the addition of mixed decimals.

Illustration. Find the total length of the metal piece shown in figure 16-2.

Solution.   5."
                      .4"
                      .3"
              + 1.2"
                6.9"

Explanation. • The numbers are placed in vertical order. Note that the figure 5 is a whole number. • Add and place decimal points as in the addition of mixed numbers.

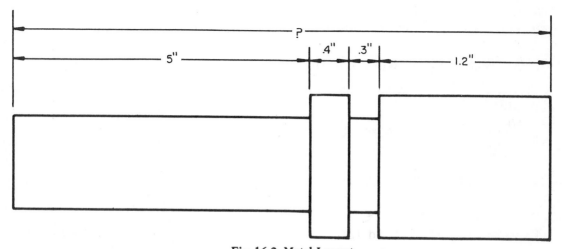

Fig. 16-2 Metal Layout

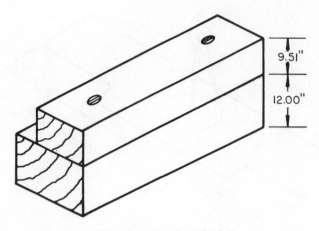

Fig. 16-3 Wooden Blocks

Illustration. Two wooden blocks are glued together as shown in figure 16-3. What is the height of the resulting piece of wood?

Solution.   9.51″
            + 12.00″
            ———
             21.51″

Explanation. • This illustration is similar to the first illustration, except that two zeros are placed after the 12 to reduce the chance of making an error in addition. • The zeros do not change the value of the number.

Dollars and fractions of dollars are added in the same manner.

Illustration. A carpenter paid $1.00 for lumber and $2.55 for a zig-zag rule. How much money did he spend?

Solution. $1.00
            + 2.55
            ———
             $3.55

Explanation. • The whole dollar numbers are placed in columns to the left of the decimal point and the fractional part of the dollar is placed to the right. The decimal point in the sum is placed directly under the decimal points in the addends.

ADDING COMMON FRACTIONS AND DECIMALS

To add common fractions and decimals, the common fractions must first be changed to decimals.

Illustration. Add 1/2, 1/4, and .8.

Solution.   1/2 = .50
            1/4 = .25
                 +.80
                 ———
                  1.55

Explanation. • The 1/2 and 1/4 are changed to decimals (.5 and .25) and placed in a vertical column. • The numbers are then added as in the preceding illustrations.

APPLICATION

1. Add each of the following pairs of whole numbers and pure decimal fractions.

   a. 6 + .70
   b. 1 + .226
   c. 8 + .41
   d. 2 + .2841
   e. 7 + .129
   f. 7 + .921
   g. 2 + .27
   h. 9 + .049
   i. 5 + .75 + 3.275 + .752
   j. 1 + .0582 + 7 + .72

   k. 8 + .059 + .101
   l. 61 + .06 + .1684
   m. .172 + 10
   n. 3 + .9056 + .56
   o. 35 + 2067 + .10056
   p. 9 + .862 + .05
   q. 546 + .677 + .93
   r. 1 + .0452 + .386
   s. 38 + .682 + .5

2. What is the outside diameter of the pipe shown in the figure below?

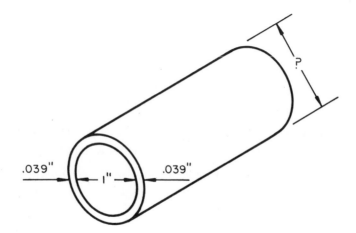

3. The rivet shown in the figure below is .125" in diameter. If another rivet is .031" larger in diameter, what is its measurement?

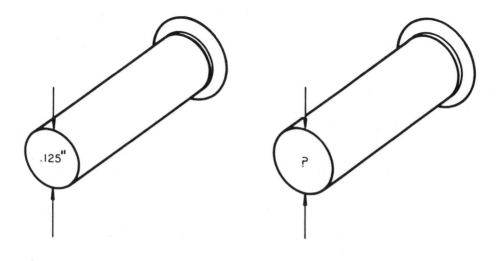

4. The length of one cast-iron pipe is 5′ and the length of another is .5′. What is the total length of the two pipes?

5. Add each of the following sets of numbers.

   a. 1/2 + 3/4 + 5/8 + 5/16   c. 1/3 + 0.335 + 0.456 + 0.0052
   b. 7/8 + 0.525 + 0.125 + 5/32   d. 0.5823 + 7/8 + 7/16 + 9/16

6. Add each of the following.

   a. 36.05 + .532 + .8 + 800 + 2000.0025
   b. 0.35 + 3.3475 + 135.6 + .02125
   c. 30.10 + 5.1 + 3.525 + 302.1205 + 18.8755
   d. 8 3/4 + .752 + 25.45 + 3.235
   e. 10.0302 + 900 + .009 + .750

7. A carpenter orders the following nails: 14 3/4 pounds of 3/4-inch brads, 15 1/5 pounds of #6d finishing, 5 pounds of #8d common, 25 1/8 pounds of #10d common, 11 7/8 pounds of #6 casing, 7 3/8 pounds of #4 box, and 13 3/5 pounds of #16d common. How many pounds does he order altogether?

8. Add the following decimals.

37.072	97.207	0.0004	640.01
+ 15.923	+ 35.003	+ 10.345	+ 5.2345

9. Three pieces of metal have the following weights: 15.5 pounds, 15.66 pounds, and 22.325 pounds. What is the total weight?

10. The thickness gauge has 8 tempered leaves of the following thickness in inches: 0.0015, 0.002, 0.003, 0.004, 0.006, 0.008, 0.10, and 0.015. What is the total thickness?

11. The measurements of a micrometer are expressed in decimals. In order to determine the dimensions of an object, addition of decimals is involved. On the sleeve of the micrometer, each exposed number is in tenths (.1) of an inch. Each uncovered subdivision indicates .025 inch. On the thimble, each of the 25 divisions indicates .001 inch.

   Using addition of decimals, find the following micrometer readings for the thickness of sheet metal.

   a. .4″ on sleeve and .012″ on thimble
   b. .5″ on sleeve and .019″ on thimble
   c. .7″ on sleeve and .008″ on thimble
   d. .3″ on sleeve and .001″ on thimble
   e. .8″ on sleeve and .018″ on thimble
   f. .6″ on sleeve and .013″ on thimble

12. A plumber cut pipes in the following lengths: 30.66″, 31.75″, 18.72″, 20.25″, 1.87″, and 9.45″. What is the total length of pipes cut?

13. Bricks are sold in units of one thousand. It takes 5.5 units of bricks for one job, 10.75 units for a second job, and 9.533 units for a third job. How many units are required for the three jobs?

14. Add the following pairs of common fractions and decimal fractions.

a. 1/4 + .0255   g. 7/8 + .355   m. 11/16 + .200   s. 15/32 + .089

b. 1/8 + .125   h. 9/16 + .878   n. 13/16 + .025   t. 13/32 + .9812

c. 2/7 + .77   i. 3/8 + .435   o. 15/16 + .031   u. 9/32 + .55556

d. 2/3 + .5   j. 3/11 + .272727   p. 19/21 + .875   v. 5/32 + .4557

e. 2/21 + .0075   k. 5/9 + .55556   q. 1/2 + .755   w. 32/59 + .542

f. 3/4 + .750   l. 8/17 + .470588   r. 1/20 + .010   x. 21/37 + .568

Unit 17
Subtraction of Decimals

OBJECTIVES

After studying this unit, the student should be able to

- subtract pure and mixed decimals.

- solve problems requiring addition and subtraction of decimals.

The processes for subtraction of decimals are similar to those for the addition of decimals. As in addition, only like quantities can be subtracted. The minuend must always be larger than the subtrahend. Decimals are sometimes carried out to four places for accuracy.

SUBTRACTING PURE DECIMALS

To subtract pure decimals, place the decimal point in the subtrahend directly below the decimal point in the minuend and subtract as usual. The decimal point in the result is placed directly below the others.

Illustration. The outside diameter of the part being measured in figure 17-1 is .9″ and the inside diameter is .4″. What is the difference in the measurements?

Solution.   .9″ minuend
     – .4″ subtrahend
     .5″ difference

Explanation. To subtract pure decimals (.9 and .4) line up the decimal points. • Then carry out the subtraction as it is done with whole numbers, and place the decimal point of the difference directly below the decimal point of the subtrahend. • The result is .5″.·

SUBTRACTING MIXED DECIMALS

In the subtraction of mixed decimals, the decimal points should be in line with each other. The whole numbers are at the left side of the decimal point and the fractions are on the right.

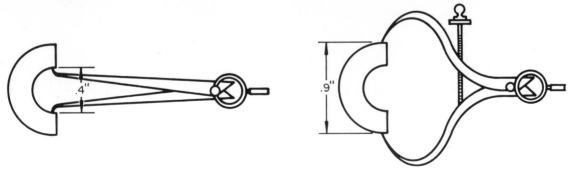

Fig. 17-1 Inside and Outside Calipers

Illustration. The thickness of one brick wall is 12.5''; the thickness of another wall is 8.25''. What is the difference in thickness of the two walls?

Solution.

$$
\begin{array}{rcl}
12.5 & = & 12.50'' \\
-\ \ 8.25 & = & -\ 8.25'' \\
\hline
& & 4.25''
\end{array}
$$

Explanation. • Before subtraction can be performed, a zero must be added to 12.5. • Subtract as usual. • Place the decimal point in the difference directly below the point in the subtrahend. • The wall measuring 12.5'' is 4.25'' thicker than the wall measuring 8.25''.

SUBTRACTING COMMON FRACTIONS AND DECIMALS

To subtract fractions from decimals, the fraction should first be changed to a decimal.

Illustration. Subtract 3/4 from .95.

Solution. Step 1.

$$3/4 = 4\overline{)3.00} \quad \begin{array}{r} .75 \\ \hline \end{array}$$

$$
\begin{array}{r}
.75 \\
4\,\overline{)3.00} \\
2\,8 \\
\hline
20 \\
20 \\
\hline
\end{array}
$$

Step 2.
$$
\begin{array}{r}
.95 \\
-\ .75 \\
\hline
.20
\end{array}
$$

Explanation. • First change 3/4 to a decimal (.75). • Then place .95 above .75 and subtract. • Be sure that the decimal point in the difference lines up with the other decimal points. • The result is .20.

SUBTRACTING DECIMALS FROM WHOLE NUMBERS

In the subtraction of decimals from whole numbers, annex the proper number of zeros after the decimal point in the whole number to carry out subtraction.

Illustration. Subtract .7581 from 9.

Solution.
$$
\begin{array}{r}
9.0000 \\
-\ \ .7581 \\
\hline
8.2419
\end{array}
$$

Explanation. • There are four figures in the subtrahend. • To subtract, four zeros must be annexed after the decimal point in the whole number. The subtraction is then carried out in the usual manner.

COMBINING THE SUBTRACTION AND ADDITION OF DECIMALS

Just as there are problems which combine addition and subtraction of whole numbers, there are problems combining addition and subtraction of decimals. When the process requires both addition and subtraction, carry out the addition first and then subtract.

Illustration. The lengths of three pieces of #22 electrical wire are 22.875', 15.752', and 9.125'. The lengths of two pieces of #24 electrical wire are 32.775' and 22.8'. What is the difference in length between the #22 wire and the #24 wire?

Solution.   Add #22 wire.   22.875'   Add #24 wire.   32.775'
                                 15.752'   + 22.800'
                               + 9.125'   55.575'
                                 47.752'

Subtract to find the difference.   55.575' #24 wire
                                 − 47.752' #22 wire
                                    7.823' difference

Explanation. The total length of the #22 wire is 47.752' and the total length of the #24 wire is 55.575'. • After obtaining the two sums, subtract to determine the difference. The #24 wire is 7.823' longer than the #22 wire.

APPLICATION

1. Subtract the following decimals.

a. .8 − .3	n. .5 − .02	aa. .65 − .56
b. .6 − .5	o. .6 − .03	ab. .86 − .57
c. .7 − .3	p. .4 − .03	ac. .25 − .19
d. .7 − .2	q. .2 − .01	ad. .56 − .39
e. .6 − .3	r. .9 − .03	ae. .37 − .024
f. .4 − .3	s. .9 − .04	af. .46 − .261
g. .2 − .1	t. .8 − .05	ag. .178 − .075
h. .9 − .4	u. .85 − .76	ah. .258 − .107
i. .9 − .3	v. .83 − .17	ai. .777 − .085
j. .8 − .5	w. .49 − .27	aj. .832 − .613
k. .8 − .03	x. .33 − .18	ak. .975 − .795
l. .6 − .05	y. .63 − .23	al. .389 − .212
m. .7 − .03	z. .97 − .86	am. .989 − .0017

2. Change each of the following pairs of numbers to decimal fractions and subtract. Round off the difference to the nearest hundredth.

a. 7/8 and 4/5	i. 2/3 and 3/4	q. 7/10 and 4/13
b. 4/5 and 5/6	j. 7/8 and 3/4	r. 13/27 and 5/81
c. 4/11 and 5/13	k. 4/7 and 7/11	s. 21/32 and 5/64
d. 7/9 and 5/13	l. 5/7 and 5/6	t. 15/16 and 15/32
e. 7/9 and 7/8	m. 9/10 and 3/7	u. 13/64 and 21/40
f. 7/8 and 11/12	n. 3/5 and 1/3	v. 9/16 and 7/12
g. 11/12 and 5/13	o. 1/20 and 3/50	w. 1/6 and 1/11
h. 13/15 and 11/12	p. 1/9 and 1/90	x. 8/9 and 11/12

3. Find the difference between 85.75 and 58.95.

4. Take 14.679 from 82.65.

5. What is the difference in the lengths of two electrical cables if one measures 75.35' and the other measures 25.72'?

6. The outside diameter of the copper tubing shown below is 1.475" and its wall thickness is .232". What is the inside diameter?

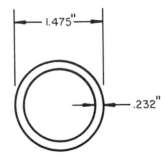

7. Subtract .8755 cubic yard from 1 cubic yard.

8. If face bricks cost $85.50 and common bricks cost $65.95, what is the difference in the price?

9. A hole .187" in diameter is needed in a piece of sheet metal. The size of the drill used is .198". Is the drill too large? If so, how much?

10. The dimensions of the template in the figure below are 1.5", 1.904", 1.879" and 1.6". The length of the template is 8.75". What is the missing dimension?

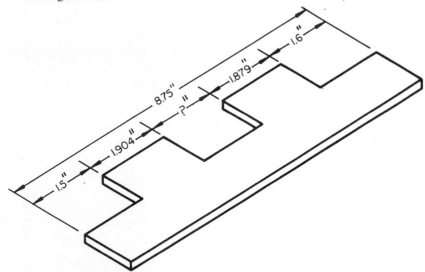

11. What is the inside diameter of a washer if the walls are .091" thick and the outside diameter measures 1.162"?

12. If the weight of the metal lath in a wall is 18.70 lb. and the weight of the lath in the ceiling is 22.10 lb., what is the difference in weight between the two laths? *Note: metal lath* is used as a plaster base. It is made from sheet metal 27" wide and 96" long. The standard weight of a metal lath is from 2.2 to 3.4 lb. per sq. yd.

13. A carpenter buys a reel containing 1500′ of electrical wire. One job requires lengths of 125.5′ and 118.750′. Another job takes lengths of 174.666′ and 236.833′. What is the difference in the total length of wire required by the two jobs?

14. Bricks used in two jobs total 500.75 and 955.25. How many bricks remain out of 2000 bricks?

15. A carpenter cuts 4 pieces of a 1/2″ water pipe which is 20.5′ long. How much is left after the pieces of the following lengths are cut: 3.5′, 3.75′, 4.25′, and 4.666′?

16. From the sum of 5/7″ and 3 1/2″, subtract the difference between 4.5″ and 5.25″.

17. Subtract the difference between 1.6 and .75 from the sum of 3/4 and 7/8.

18. A box contains 75.528 lb. of #6d finishing nails, from which 22.929 lb. are used. How many pounds are left?

19. The length of a room on a floor plan is 15.782′ and the width is 11.998′. What is the difference in dimensions?

20. As shown in the figure below, the length of a standard face brick is 7 5/8″ and the length of a standard concrete block is 15.625″. What is the difference in their lengths?

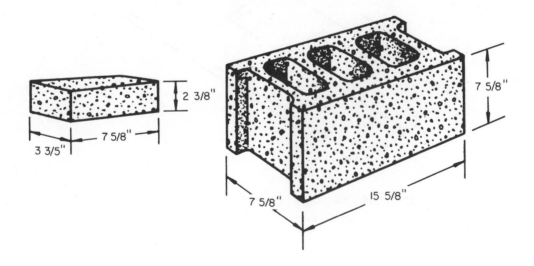

21. The height of a standard brick is 2.375″ and the height of a standard concrete block is 7.625″. What is the difference in height?

22. A square yard is an area equivalent to 3′ x 3′, or 9 square feet. (The thickness is not considered.) A house has wall-to-wall carpet which is placed in the living room and bedroom. In the bedroom there are 31.6 square yards and in the living room 26.16 square yards. What is the difference in the number of square yards required for each room?

23. Draw a line 5.25″ long and then draw one .750″ shorter. How long is the second line?

24. The floor of a structure is 8.083′ from the ceiling, and a window top is 1.5′ from the ceiling. What is the distance from the window top to the floor?

Unit 18
Multiplication of Decimals

OBJECTIVES

After studying this unit, the student should be able to

- multiply decimals.
- determine the length of common, hip, and valley rafters.
- figure costs of materials.

Methods for multiplying decimals are similar to those for multiplying whole numbers except for the location of decimal points. Decimal points in multiplication problems are not lined up as in addition and subtraction. Their location is determined by the sum of the decimal places in the multiplicand and multiplier.

MULTIPLYING PURE DECIMALS

When multiplying decimals, the multiplicand is multiplied by the multiplier. To place the decimal point, begin at the right of the product and point off as many decimal places as there are in both the multiplicand and multiplier. Add as many zeros as are necessary to obtain the required number of decimal places in the product.

Illustration. Multiply .55 by .23.

Solution.
```
     .55
   x .23
     165
     110
    .1265
```

Explanation. Multiply as whole numbers are multiplied. There are two decimal places in the multiplicand and two decimal places in the multiplier, so four decimal places should appear in the product. The product is .1265.

MULTIPLYING MIXED DECIMALS

Problems involving multiplication of mixed decimals frequently occur in carpentry work.

Illustration. How many square inches are there in a piece of sheet metal 5.251″ x 6.522″?

Solution.
```
       5.251″
     x 6.522″
       10502
       10502
       26255
      31506
     34.247022 sq. in.
```

Explanation. • Multiply as whole numbers are multiplied. • There are three decimal places in the multiplicand and three places in the multiplier. • Therefore, there are six decimal places in the product: 34.247022 sq. in.

MULTIPLYING DECIMALS BY WHOLE NUMBERS

Cost is often determined by multiplying decimals by whole numbers.

Illustration. If a sheet of copper costs \$.52 per pound, how much does 79 lb. cost?

Solution.

```
 $  .52
 x  79
   468
  364
$41.08
```

Explanation. • Multiply 79 lb. by the cost for each pound (\$.52). • There are two decimal places in the multiplicand and none in the multiplier; therefore, point off two places in the product. • The answer is \$41.08.

PRACTICAL APPLICATION: DETERMINING THE LENGTH OF COMMON RAFTERS

A *common rafter* is that roof member which extends from the rafter plate to the ridge of the roof, figure 18-1. The length of a common rafter is measured directly from end to end by using the rafter table which appears on a framing square, figure 18-2. The length of the common rafter per foot of run is found on the face of the square. *Note:* A *framing square* is a right angle tool used to measure material for structural framing.

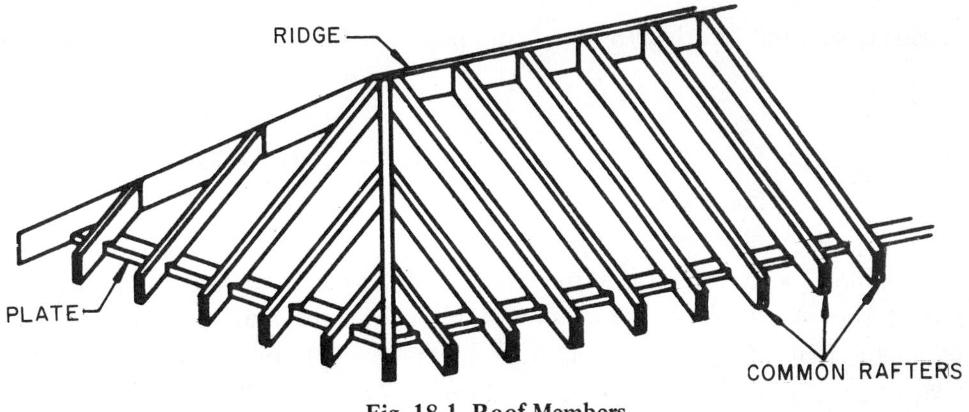
RIDGE
PLATE
COMMON RAFTERS

Fig. 18-1 Roof Members

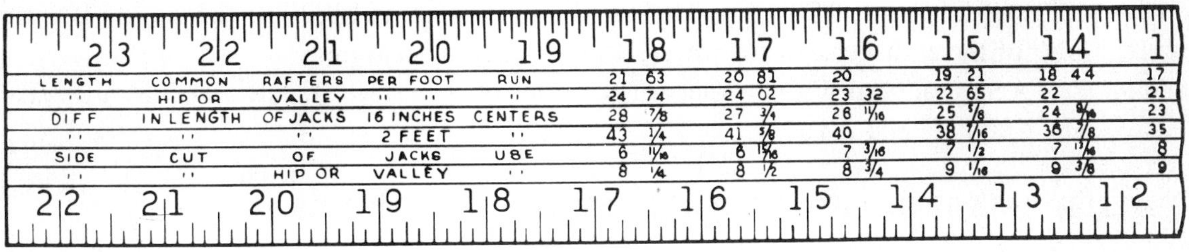
23 22 21 20 19 18 17 16 15 14 1
LENGTH COMMON RAFTERS PER FOOT RUN 21 63 20 81 20 19 21 18 44 17
" HIP OR VALLEY " " " 24 74 24 02 23 32 22 65 22 21
DIFF IN LENGTH OF JACKS 16 INCHES CENTERS 28 7/8 27 3/4 26 11/16 25 5/8 24 9/16 23
" " " 2 FEET " 43 1/4 41 5/8 40 38 7/16 36 7/8 35
SIDE CUT OF JACKS USE 6 11/16 6 15/16 7 3/16 7 1/2 7 13/16 8
" " HIP OR VALLEY " 8 1/4 8 1/2 8 3/4 9 1/16 9 3/8 9
22 21 20 19 18 17 16 15 14 13 12

Fig. 18-2 Rafter Table (Steel Square)

Rise Per Foot of Run	Length of Common Rafter Per Foot of Run	Length of Hip or Valley Rafter Per Foot of Run	Difference in Length of Jack (16" O.C.)	Difference in Length of Jack (24" O.C.)
18"	21.63	24.74	2' 4 7/8"	3' 7 1/4"
17"	20.81	24.02	2' 3 3/4"	3' 5 5/8"
16"	20.00	23.32	2' 2 11/16"	3' 4"
15"	19.21	22.65	2' 1 5/8"	3' 2 7/16"
14"	18.44	22.00	2' 0 9/16"	3' 0 7/8"
13"	17.69	21.38	23 9/16"	2' 11 3/8"
12"	16.97	20.78	22 5/8"	2' 9 15/16"
11"	16.28	20.22	21 11/16"	2' 8 9/16"
10"	15.62	19.70	20 13/16"	2' 7 1/4"
9"	15.00	19.21	20"	2' 6"
8"	14.42	18.76	19 1/4"	2' 4 7/8"
7"	13.89	18.36	18 1/2"	2' 3 3/4"
6"	13.42	18.00	17 7/8"	2' 2 13/16"
5"	13.00	17.69	17 5/16"	2' 2"
4"	12.65	17.44	16 7/8"	2' 1 5/16"
3"	12.37	17.23	16 1/2"	2' 0 3/4"

Fig. 18-3 Rafter Table, Steel Square

A more detailed table is shown in figure 18-3.

To determine the length of a common rafter, multiply the length per foot of run given on the framing square table by the actual number of feet of the span. The tail of the rafter is added to the length.

Illustration. Find the length of a common rafter for a span of 24' and 1/4 pitch.

Solution.

$$13.42''$$
$$\times \quad 12$$
$$\overline{2684}$$
$$\underline{1342}$$
$$161.04''$$

$161.04'' \div 12 = 13.42'$
$.42' \times 12 = 5.04''$
$13.5'' - 13/16'' = 13' 4\ 13/16''$

Explanation. • By using the rafter table, find the length of a common rafter when the rise of roof is 6" per foot run, or 1/4 pitch, and the building span is 24'. Directly under the figure 6 is 13.42, which represents the length of the rafter in inches for each foot of run for a 1/4 pitch. • The building is 24' wide. The run is half of 24, or 12'. • Multiply 13.42" by 12'. • The product is 161.04". • To change 161.04" to feet, divide 161.04" by 12", which equals 13.42'. • To change .42' to inches, multiply by 12". • The result is 5 1/25". The length of the rafter is 13'5". If the answer is within 1/8" of the actual size of the rafter, it is considered accurate.

• Subtract half the thickness of the ridge board (1 5/8"); half of 1 5/8" is 13/16". The result is 13' 4 3/16".

DETERMINING THE LENGTH OF HIP AND VALLEY RAFTERS

A *hip rafter* is the roof member which extends diagonally from the corner of the building to the ridge and forms a hip on the corner of a building. The *valley rafter* is similar to the hip rafter except that it forms a valley instead of a hip. The hip and valley rafter lengths are the same for the run on a roof.

The length of a hip and valley rafter can be determined from a table on the framing square, just as the length of common rafters can be found.

Their lengths are found on the second line of the rafter table on the framing square labeled "Length of hip or valley rafters per foot run" (of common rafters). Multiply the length per foot of run found in the table by the actual run of the common rafters.

Illustration. Find the length of a hip or valley rafter for a roof with a span of 18′ and 1/3 pitch.

Solution.   18.76″ x 9 = 168.84″

$$\frac{168.84''}{12''} = 14.07' = 14' \ 1/16''$$

Explanation. • The 1/3 pitch indicates that the rafter rises 8″ for each foot of run. • The 18.76, the length of hip or valley rafters in inches, is found on the second line under the figure 8. • The run of the common rafter is 9′.
    • Multiply 18.76″ by 9; the product is 168.84″. • Change 168.84″ to feet. • The final answer is 14.07′ or 14′ 1/16″.

DETERMINING THE COST OF LUMBER

When determining the cost of lumber, the unit of measurement is considered to be 1000 bd. ft. To determine the total cost, multiply the cost of each board foot by the number of board feet.

Illustration. A certain grade of lumber costs $85.45 per 1,000 bd. ft. Find the cost of 742 bd. ft.

Solution. $\frac{\$85.45}{1000} = \$.08545$

    $.08545 x 742 = 63.40390 or $63.40

Explanation. • Find the cost of each board foot ($85.45 ÷ 1000 = $.08545). • Multiply $.08545 by 742 ($63.40390) to find the total cost. • Rounded off, the answer is $63.40.

MULTIPLYING DECIMALS BY FRACTIONS

Illustration. Multiply .24 by 2/3.

Solution.   .24 x 2/3 = .48/3 = .16

Explanation. • Multiply the multiplicand (.24) by the numerator of the fraction 2/3 (2). • The product (.48) becomes the numerator of a new fraction (.48/3). • Divide .48 by 3. The result is .16.

APPLICATION

1. Multiply the following.

a. 45
   .03
	e. 3.60
	   .12
		i. 4.29
		  5/6
			m. 625
			   .49

b. \$.24
  3/4
	f. \$22.35
	  .14 2/5
		j. 1.28
		  .06
			n. 2.3 1/2
			1.3 3/4

c. \$38.75
  .15 2/3
	g. .04
	  .05
		k. 3.43
		  .57
			o. 5.2 3/4
			  .2 7/8

d. .2315
  2
	h. 3.60
	1/12
		l. 3.91
		  .03 1/6	

2. A sheet of metal is .6′ wide and .8′ long. How many square inches are there in the sheet?

3. A sheet metal worker is constructing a water gutter from a piece of metal which is 20′ long and .8′ wide. How many square feet are there in the metal piece?

4. If copper tubing costs \$.23 per foot, how much does 25.628′ cost?

5. The price of a type of electrical wire is \$.09 per foot. How much does a 250.75′ piece cost?

6. A type of nail costs \$.23 1/2 per pound. What is the cost of .53 3/4 pounds?

7. Find the cost of each of the following items.

   a. 5 insulators @ \$.08 each
   b. .93 lb. solder @ \$.83 each
   c. 3.4 lb. of chalk @ \$.21 lb.
   d. 6 bd. ft. of lumber @ \$.91 each
   e. 2.5 boxes of staples @ \$.75 per box

8. Find the run of a roof with a 10′ span and equal pitch.

9. Determine the length of the hip and common rafters for a roof with a 1/3 pitch. The span is 28′.

10. What is the length of the main rafter per foot run for a roof with a 1/6 pitch?

11. What is the length per foot of run for a hip or valley rafter on a roof with a 1/6 pitch?

12. What is the length of a main rafter per foot run for a roof with a 1/2 pitch?

13. What is the length of a hip or valley rafter per foot run for a roof with 1/2 pitch?

14. A grade of red oak lumber costs \$110.00 per 1,000 board feet. Find the cost of 1250 board feet.

15. A roll of copper tubing weighs 9.41 lb. Find the cost of the roll if it sells for \$1.12 per pound.

Unit 19
Division of Decimals

OBJECTIVES

After studying this unit, the student should be able to

- divide whole numbers by decimals and decimals by whole numbers.
- divide decimals by decimals and use short cuts when dividing decimals.
- determine the pitch of a roof.

The division of decimals is similar to the division of whole numbers. The dividend is placed inside a division bracket and the divisor is placed outside the bracket. If the divisor is a decimal, the decimal point is moved to the right, thus making the divisor a whole number. The decimal point in the dividend is moved to the right the same number of places. If the dividend has fewer digits than the divisor, zeros are annexed to the dividend until both the dividend and divisor have the same number of digits.

DIVIDING WHOLE NUMBERS BY DECIMALS

Illustration. How many pieces measuring .25″ can be cut from a strip of metal 14″ long as shown in figure 19-1?

Solution. Divide 14″ by .25″.

$$.25. \overline{)\,14.00.} \quad \begin{array}{r} 56. \\ \underline{125} \\ 150 \\ \underline{150} \end{array}$$

Explanation. • In this illustration two zeros are added to the dividend to make the number of decimal places equal to those in the divisor. • The decimal point in the divisor is moved to the far right (two places); therefore, the decimal point in the dividend is moved two places with the decimal point in the quotient above it.

• Sometimes a zero must be annexed to the quotient.

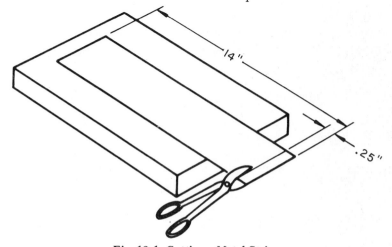

Fig. 19-1 Cutting a Metal Strip

DIVIDING DECIMALS BY WHOLE NUMBERS

Illustration. The results of micrometer measurements are always expressed as decimals. Assume that the reading on a micrometer indicates a space .092″ thick. The space should be filled with four pieces of material. What is the thickness of each piece of material?

Solution.   Divide .092 by 4.

$$
\begin{array}{r}
.023 \\
4\overline{)\ .092} \\
\underline{8} \\
12 \\
\underline{12} \\
\end{array}
$$

Explanation. • Divide as in division of whole numbers, applying rules for division of decimals. Notice that the divisor does not contain a decimal point; therefore, the decimal point in the dividend is not moved to the right. The decimal point in the quotient is directly in line with the decimal point in the dividend.

DIVIDING DECIMALS BY DECIMALS

Remember that the decimal point determines the value of the quotient. This can be illustrated by the division of a decimal by another decimal.

Illustration. The piece of metal in figure 19-2 is 8.025″ long. How many .03″ pieces can be cut? Do not consider the waste.

Solution.   Divide 8.025 by .03.

$$
\begin{array}{r}
267.5 \\
.03.\overline{)\ 8.02.5} \\
\underline{6} \\
20 \\
\underline{18} \\
22 \\
\underline{21} \\
15 \\
\underline{15} \\
\end{array}
$$

Explanation. • Move the decimal point in the divisor (.03) to the extreme right (two places) to form a new number (03.). • Move the decimal point in the dividend (8.025) two places. • Position the decimal point in the quotient and divide as usual.

Fig. 19-2 Sawing Metal on a Power Saw

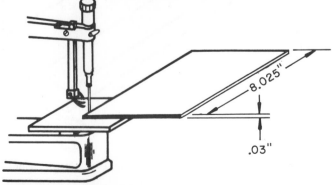

SHORTCUTS IN DIVIDING DECIMALS

To divide by 10, 100, or 1000 move the decimal point in the quotient one place to the left for each zero in the divisor.

Illustration. Divide 16.424 by 10.

Divide 17.635 by 100.

Divide 18.876 by 1000.

Solution. $16.424 \div 10 = 1.6424$

$17.635 \div 100 = .17635$

$18.876 \div 1000 = .018876$

Explanation. • By changing the position of the decimal point, the value of the fraction changes. • Notice that in some cases, zeros must be annexed in order to divide.

APPLICATION

1. Divide the following decimal fractions. Carry out each answer to at least one decimal place.

   a. $82.112 \div 75$
   b. $173.375 \div .489$
   c. $.014 \div .024$
   d. $.4 \div .003$
   e. $093.5 \div 27.8$
   f. $60.2 \div 1.9$

   g. $1 \div 0.78554$
   h. $25.11 \div 81$
   i. $.003 \div .0625$
   j. $310 \div .0625$
   k. $4.80 \div 8.4$

2. Use the shortcut method to divide the following numbers.

   a. $18.244 \div 10$
   b. $32.612 \div 100$

   c. $44.125 \div 1000$

3. A bolt measuring 15/16″ x 1.5″ weighs 0.375 lb. How many bolts are there in 75 lb.?

4. A group of 16 carpenters receives a bonus of $975.70 which is to be divided equally among them. How much should each person receive?

5. A plumber connects four building sewers to the public sewers shown below. The total bill is $226.32. What is the average cost for each sewer?

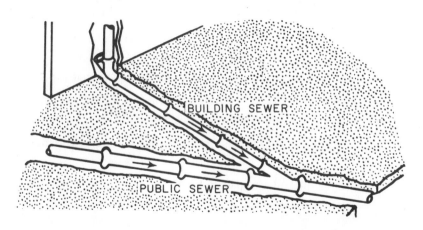

6. If a dozen roundhead metal screws purchased separately cost $4.80 and a dozen purchased in a lot cost $4.20, how much money is saved per screw by purchasing by the dozen?

7. If 21 rolls of solder weigh 153.93 lb., how much does 1 roll weigh?

8. A carpenter works 189.75 hours in one month. How many hours does he work each day if the month contains 23 workdays?

9. How long will 55.25 lb. of lead last if 3.25 lb. are used each day?

10. If a carpenter nails 836.2 sq. ft. of drop siding, shown below, on a building in 45.2 hours, how many square feet does he nail in 1 hour?

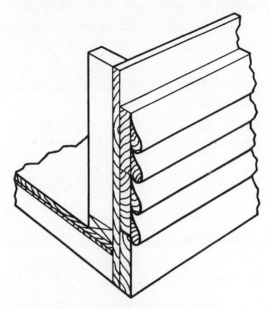

11. The length of a common rafter is 203.64″. How many feet is this?

12. The length of the hip rafter is 180.47″. How many feet is this?

13. If a valley rafter measures 244.69″ long, what is the measurement in feet?

14. There are 7500 bricks in a brick wall, with 7.5 bricks to each square foot. What is the size of the wall in square feet?

15. How many pieces 2.5′ long can be cut from a 5′ pipe?

16. How many pieces 2.5′ long can be cut from a galvanized pipe 20′ long?

17. The length of a piece of #14 electrical wire is 93.5′. How many pieces 5.5′ long can be cut from it?

18. Rafters are spaced 1.5′ O.C. on a building 67.5′ long. How many equal spaces are there?

19. Studs are to be placed on a wall 19.95′ long spaced 1.33′ O.C. How many studs are needed?

20. Ceiling joists, shown below, are to be placed on a building 37.6' long, spaced 1.83' O.C. How many joists are needed?

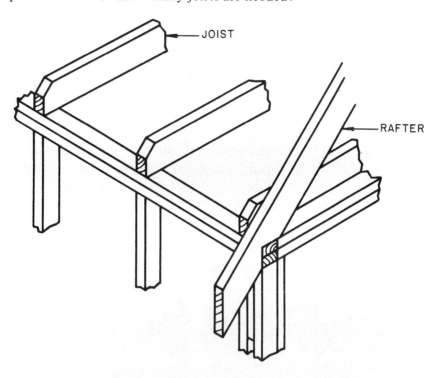

CAREER PROFILE: THE ROUGH FRAMER

Job Description

Rough framers begin their work after the foundation is constructed. They determine the methods and sequence of proposed construction of foundation sills, girders, columns, joists, subfloors, and other elements of a building frame. They read blueprints and specifications to install plaster grounds and lath. The framer generally works outdoors. In cold, wet climates, work may be seasonal.

Qualifications

Carpenters who work with rough framing should have special courses in blueprint reading and industrial materials. A sound background in mathematics and sciences is especially useful. Good physical health and the ability to work in high places are necessary. A cooperative attitude is very important, since the framer usually works within a group of construction workers.

Section 3 Ratio, Proportion, and Percent

Unit 20
Ratio, Proportion, and Averages

OBJECTIVES

After studying this unit, the student should be able to

- determine the relationship of one number to another by using ratio.
- find an unknown term in a proportion when three quantities are given.
- use direct and inverse proportion.
- determine the average of two or more quantities.

RATIO

Ratio is used to express the relationship between two like quantities. The relationship is determined by dividing the first quantity, or *antecedent,* and the second quantity, or *consequent,* by the same factor. The two quantities being compared are known as the *terms* of the ratio. The ratio may be expressed as a fraction. *Note:* Only *like* quantities can be expressed in a ratio; 9′ cannot be compared to 3″ unless inches are converted to feet or feet to inches.

The sign : is often used to indicate a ratio. The expression 8/12, or 8 ÷ 12, may be written as the ratio 8:12. The ratio is calculated by dividing 8 and 12 by 4, which equals 2/3. The ratio of one number to another represents its value as compared with the other.

Illustration. What is the actual reduced ratio of gears in the portable electric saw in figure 20-1 when its ratio is expressed as 6 to 2?

Solution. 6/2 = 3/1 or 3 to 1.

Explanation. • In this case, 6 is divided by 2, which equals 3. When 6:2 is reduced, it becomes 3:1. When one gear makes 6 complete revolutions, the other gear makes 2 or, when one makes 3 revolutions, the other makes 1. Therefore, the ratio can be expressed as 3:1 or 6:2.

Fig. 20-1 Power Saw

113

PROPORTION

A *proportion* is an expression of equality between two ratios. An equal sign (=) or the symbol :: is used in the proportion. For example, the proportion 2:10 = 4:20 shows that the two ratios are equal to one another. The proportion could also be expressed as 2/10 = 4/20. The first and last terms of the proportion are called the *extremes* (in this case, 2 and 20), and the two middle terms are called the *means* (10 and 4).

In a proportion, *the product of the means is equal to the product of the extremes.*

Illustration. Show that the proportion 6:2::3:1 is true.

Solution.   6 x 1 = 6   2 x 3 = 6

Explanation. • The product of the extremes (6 and 1) is equal to the product of the means (2 and 3), so the expressed proportion is correct.

The proportion can also be expressed as 6/2 = 3/1, which is stated "6 is to 2 as 3 is to 1."

FINDING AN UNKNOWN FACTOR

When one term in a proportion is unknown and the other three are known, the unknown term can be found by multiplication and division.

Illustration. The proportion illustrated in figure 20-2 may be expressed as 9:3::3:e. Find the unknown term.

Solution.   9:3::3:e   *Check:*   9:3::3:1
Step 1. 9 x e = 3 x 3   9 x 1 = 9
Step 2. (3 x 3) ÷ 9 = e   3 x 3 = 9
Step 3. e = 9/9 = 1

Explanation. • The letter e represents the unknown extreme. • Since the product of the extremes must equal the product of the means, the unknown extreme can be found by dividing the product of the means by the extreme. • The result is 1. • The proportion is then expressed as 9:3::3:1.

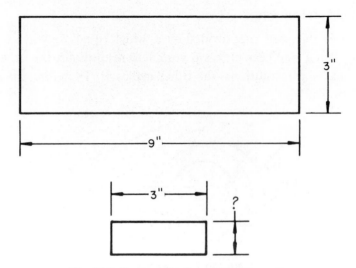

Fig. 20-2 Rectangular Layout in Proportion

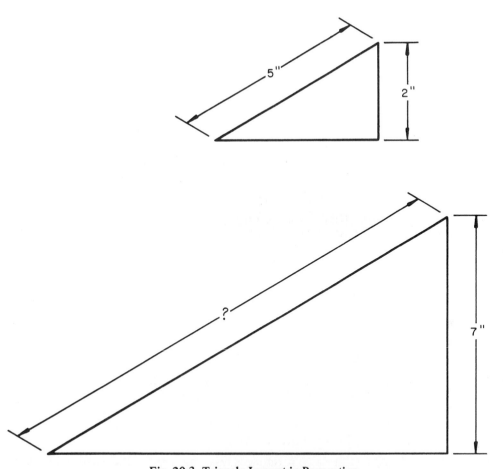

Fig. 20-3 Triangle Layout in Proportion

An unknown mean in a proportion may be found by the same method.

Illustration. Refer to figure 20-3. Find the unknown as expressed in the proportion 5:2::m:7.

Solution. 5:2::m:7
Step 1. 5 x 7 = 2 x m
Step 2. 5 x 7 = 35
Step 3. 35 ÷ 2 = m
m = 17 1/2

Explanation. The letter *m* represents the unknown mean. In this illustration the product of the extremes (5 and 7) is divided by the known mean. The result is 17 1/2.

Direct Proportion

Proportion covered so far may be defined as *direct* proportion. In direct proportion, as the terms of one ratio increase or decrease, the terms of the other ratio also increase or decrease in proportion.

Illustration. A carpenter needs to know how many stop-and-waste valves can be bought with $12.00 if 8 valves cost $16.00.

Solution.   16:12::8:e
             Step 1. 16 x e = 16e
             Step 2. 12 x 8 = 96
             Step 3. e = 96/16 = 6
                  16:12::8:6

Check   16 x 6 = 96
            12 x 8 = 96

Explanation. • In this proportion, the figures representing the dollars are in ratio. • The figures representing the valves (12 and 8) are in ratio. • The illustration shows that since $16.00 buys 8 valves, 6 valves may be purchased with $12.00. • Following the rules of direct proportion, as the dollars in the problem increase, the number of valves that may be purchased also increases.

Indirect Proportion

In *indirect,* or *inverse* proportion, two of the terms in one of the ratios are reversed.

Illustration. If 5 carpenters can install paneling in 7 days, how many days are required for 8 carpenters to perform the same job? Assume that they are working at the same speed.

Solution.   5:8::m:7   *Check* 5:8::4 3/8:7
              Step 1. 5 x 7 = 35   5 x 7 = 35
              Step 2. 8 x m = 8m   8 x 4 3/8 = 35
              Step 3. m = 35/8 = 4 3/8

Explanation. • Since in this case, it requires less time for the 8 carpenters to complete the job than the 5 carpenters, indirect proportion is used. • Instead of writing the proportion 5:8::7:e, it is written with two of the terms reversed: 5:8::m:7. • Multiply 5 x 7 (35), and 8 x m (8m). • Then divide 35 by 8 to find the missing term, 4 3/8.

Ratios which make up proportions may be expressed in fractional form. For example, the indirect proportion written as 5:8::m:7 may also be written as 5/8 = m/7.

AVERAGES

Determining averages requires both addition and division.

To determine the average of two or more quantities, divide the sum of the quantities by the number of quantities.

This rule may be applied when finding the average of whole numbers, fractions, mixed numbers, and decimal fractions.

Illustration. A carpenter is installing 6 pieces of electrical wire, figure 20-4. What is the average length of one piece?

Solution.   12' + 8' + 10' + 5' + 6' + 7' = 48'
             48' ÷ 6 = 8

Explanation. • Add the figures to find the total number of feet (48). • Count the number of quantities involved (in this case, 6, the number of pieces of wire). Divide

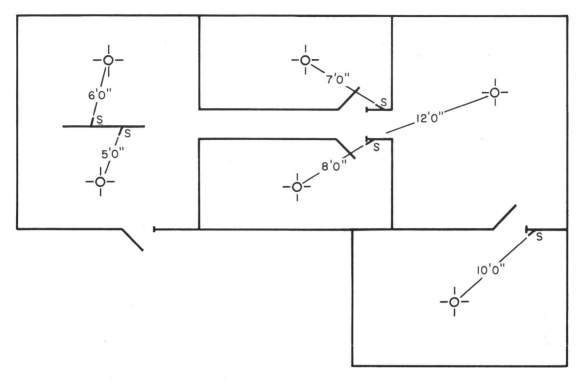

Fig. 20-4 Electrical Wire Length

the total number of feet by the number of pieces of wire to determine the average length of a piece of wire, 8.

Illustration. Find the average width of 4 dowels which measure 1/2″, 5/8″, 3/4″, and 3/16″.

Solution.   1/2″ = 8/16″
5/8″ = 10/16″
3/4″ = 12/16″
3/16″ = 3/16″
        33/16 ÷ 4 = 33/16 x 1/4 = 33/64″

Explanation. • Find the lowest common denominator and find the total of the numerators (33). • Divide by the number of dowels (4). • The average width of the dowels is 33/64″.

Illustration. Five measurements are taken on the micrometer in figure 20-5, page 118; .003, .031, .175, .256, and .315. What is the average reading?

Solution.   .003
.031
.175
.256
.315
.780   .780 ÷ 5 = .156

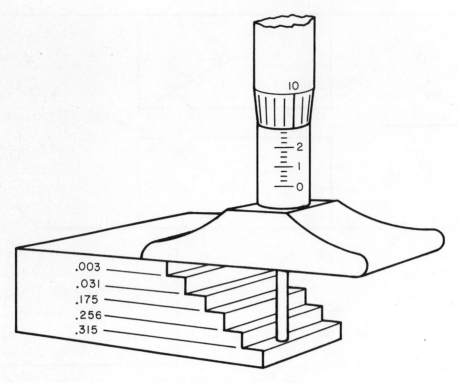

Fig. 20-5 Decimal Readings

Explanation. • Add the decimal fractions. • Then divide the sum (.780) by the number of decimals (5). • The average micrometer reading is .156.

APPLICATION

1. What is the ratio of 3 to 5 expressed as a fraction?

2. Express the ratio of 3 to 5 using the proportional sign.

3. What is the actual gear ratio when its ratio is expressed as 6 to 2?

4. Find the actual ratio of each expression.

a. 9 to 3	f. 7 to 4	k. 14 to 21	p. 5 to 10
b. 6 to 8	g. 8 to 15	l. 2 to 8	q. 12 to 6
c. 16 to 20	h. 2 to 12	m. 9 to 6	r. 3 to 4
d. 8 to 2	i. 12 to 3	n. 4 to 3	s. 2 to 9
e. 6 to 11	j. 8 to 6	o. 4 to 24	t. 10 to 5

5. Find the unknown term in each of the following expressions.

a. 3:2::2:?	d. 3:1::2:?	g. 16:20::?:21	i. 6:3::8:?
b. 25:5::35:?	e. 10:5::?:15	h. 20:4::?:20	j. 25:5::?:35
c. 30:6::?:40	f. 5:3::6:?		

6. How many cubic yards of concrete can be purchased with $32.50 if 5 cu. yd. cost $81.25?

7. Find the average of the following groups of mixed numbers.

a. 1 1/2	d. 2 3/64	f. 2 1/3	h. 10 13/64
2 1/4	2 3/32	2 2/5	3 5/32
3 3/8	3 11/16	3 1/9	4 5/16
b. 3 1/16	e. 2 1/2	g. 12 1/6	i. 2 5/15
3 3/4	3 3/16	5 5/9	3 1/6
2 5/8	3 3/32	3 2/3	4 2/3
c. 3 1/32			
2 3/16			
2 7/8			

8. Find the average of the following groups of fractions.

   a. 1/2, 1/7, 5/8   d. 3/7, 2/7, 5/7
   b. 3/4, 3/9, 4/5   e. 3/64, 9/64, 11/64
   c. 7/8, 5/6, 2/3

9. Find the average length of galvanized pipes which measure 4' 6'', 15' 7'', and 7' 11''.

10. The inside diameters of the pipes in figure 20-6 measure 1/4'', 3/8'', 1/2'' and 9/16''. What does the average inside diameter measure?

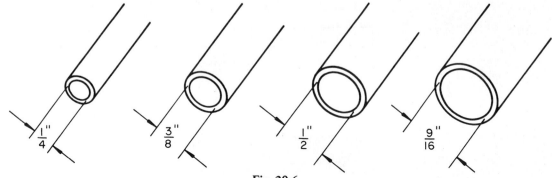

Fig. 20-6

11. A contractor builds 5 houses which cost $12,800, $8,900, $10,150, $11,555, and $15,655. What is the average cost per house?

12. In 5 workdays, a carpenter earns $45.00, $50.00, $20.00, $55.00, and $60.00. What is the average earning per day?

13. A carpenter saws 4 studs which measure 8 1/2', 9 1/4', 7 2/3', and 8 3/8'. Find the average length per stud.

Unit 21
Determining Percents and Quantity

OBJECTIVES

After studying this unit, the student should be able to

- express quantities in percents.
- change percents to decimals, percents to fractions, and fractions to percents.
- compare numbers by using percents.
- determine a quantity or number with a given percent.

A percent is a fractional part of a whole expressed in hundredths. The percent sign (%) is used to indicate the number of hundredths expressed.

The decimals .01, .05, and .25 are read as one hundredth (1%), five hundredths (5%), and twenty-five hundredths (25%). One hundredth is one part of 100 parts, five hundredths is 5 parts of 100 parts, and twenty-five hundredths is 25 parts of 100 parts. The whole of any number contains 100% of itself.

Figure 21-1 illustrates different percents of a whole. Each small square represents 1% (.01 or 1/100) of the total quantity.

Figure 21-2 shows some quantities expressed in different terms.

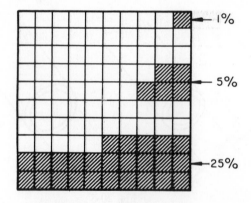

Fig. 21-1 Percents

Proper Fractions	Decimal Hundredth	Percent	Fractional Hundredth	Proper Fractions	Decimal Hundredth	Percent	Fractional Hundredth
1/20	.05	5%	5/100	1/2	.50	50%	50/100
1/10	.10	10%	10/100	3/5	.60	60%	60/100
1/8	.12 1/2	12 1/2%	12 1/2/100	2/3	.66 2/3	66 2/3%	66 2/3/100
1/6	.16 2/3	16 2/3%	16 2/3/100	7/10	.70	70%	70/100
1/5	.20	20%	20/100	3/4	.75	75%	75/100
1/4	.25	25%	25/100	4/5	.80	80%	80/100
1/3	.33 1/3	33 1/3%	33 1/3/100	7/8	.87 1/2	87 1/2%	87 1/2/100
2/5	.40	40%	40/100	9/10	.90	90%	90/100

Fig. 21-2 Fractional, Decimal, and Percent Equivalents

When working with percents, remember that a whole is 100 parts or 100% of itself. Any percent greater than 100% represents a whole and a fractional part of another whole.

CHANGING PERCENTS TO DECIMALS

Before a percent can be used in problems, it must be expressed as a decimal or fraction. First, the percent sign is dropped. The decimal point is then moved to the left two places, which is dividing the number by 100.

Illustration. Change 1%, 100%, 275%, and 10% to decimals.

Solution.
$$1\% = .01$$
$$100\% = 1.00$$
$$275\% = 2.75$$
$$10\% = .10$$

Explanation. • When 1% is changed to a decimal, drop the percent sign and divide by 100. A zero is prefixed to the quotient so that there are enough places to point off two places. • Notice that there are three digits in 100%. Drop the percent sign and place a decimal point two places to the left. • The result of dividing 275 by 100 is 2.75, showing that 275% is greater than 1.

CHANGING PERCENTS TO FRACTIONS

A percent may be changed to a decimal before being changed to a fraction, or reduced directly to a fraction.

Illustration. Change 6%, 12%, 300%, and 216% to decimals, to fractions, and then to whole numbers.

Solution.
$$6\% = .06 = 6/100 = 3/50$$
$$12\% = .12 = 12/100 = 3/25$$
$$300\% = 3.0 = 300/100 = 3/1 = 3$$
$$216\% = 2.16 = 2\ 16/100 = 2\ 4/25$$

Explanation. • To change a decimal fraction to a common fraction, express the decimal as a common fraction with the appropriate denominator. • Reduce the fraction to lowest terms.

To directly reduce the percent to a fraction, drop the percent sign. Express the percent as the numerator of a fraction with 100 as the denominator. Reduce to lowest terms.

Illustration. Change 5%, 15%, 100%, 1/3%, and 116 2/3% to common fractions, whole numbers, or mixed numbers.

Solution.
$$5\% = 5/100 = 1/20$$
$$15\% = 15/100 = 3/20$$
$$100\% = 100/100 = 1$$
$$1/3\% = \frac{1/3}{100} = 1/3 \div 100 = 1/3 \times 1/100 = 1/300$$
$$116\ 2/3\% = \frac{116\ 2/3}{100} = 116\ 2/3 \div 100 = 350/3 \times 1/100 = 350/300 = 1\ 1/6$$

Explanation. • Since *percent* (%) stands for hundredths, each fraction has 100 as a denominator. • The percents 5%, 15%, and 100% are changed by dropping the percent sign, placing those numbers as numerators and 100 as the denominator, and reducing the fractions to lowest terms. • Notice that a fraction or mixed number is used as a numerator in the problems involving 1/3% and 116 2/3%. • The fraction line indicates division, so the numerator can be divided by the denominator. • The answer is then reduced to lowest terms.

CHANGING A FRACTION TO PERCENT

To change a fraction to a percent, first change the fraction to a decimal and then write it as a percent.

Illustration. Change 3/4 and 1 1/2 to percents.

Solution. 3/4 = 3 ÷ 4 = .75 = 75%
1 1/2 = 3/2 = 3 ÷ 2 = 1.5 = 150%

Explanation. • When the fraction is a common fraction, change it to a decimal by dividing the numerator (3) by the denominator (4). • The result is .75. • Multiply by 100 by moving the decimal point 2 places to the right and annex the percent sign.

When the fraction is a mixed number such as 1 1/2, change it to an improper fraction (3/2) and then to a decimal (1.5). Multiply by 100 and annex the percent sign.

DETERMINING PERCENTAGES

There are three elements involved in percentage problems: the *base,* the *rate,* and the *percentage.*

Persons in the carpentry business are aware of the importance of using material so that there is as little waste as possible. In carpentry, the *waste* is the excess material cut away from a given project while completing the job. The amount of waste, expressed as a percentage, varies according to the way in which the material is laid out, cut, and installed. For example, in figure 21-3, there is less waste if the sheathing is installed horizontally rather than diagonally.

To find a percentage of any quantity, change the rate to decimal hundredths and multiply by the base. Point off in the product as many decimal places as there are in the multiplicand and the multiplier combined.

Illustration. The plan in figure 21-4 shows that 600 sq. ft. of material is needed to install subflooring. It is recommended that 20% be added for waste. What is the amount of waste material?

Solution. 20% = .20   600 base
.20 rate
120.00 percentage

Explanation. • Change 20% to .20 and multiply by 600 sq. ft. • The amount of waste called for is 120 sq. ft.

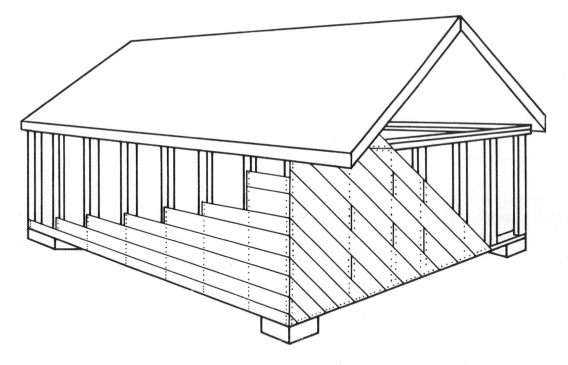

Fig. 21-3 Horizontal and Diagonal Sheathing

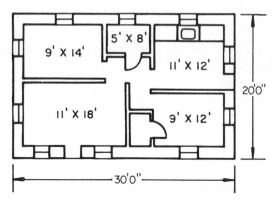

Fig. 21-4 Floor Plan

In figuring the board feet of flooring needed to cover a given area, find the area in square feet. Add that figure to the percentage of waste allowed for matching, cutting, and shrinkage.

Illustration. A floor measures 250 sq. ft. The waste allowed is 42% of this figure. What is the amount of waste in square feet?

Solution.   42% = .42   250 sq. ft.
    .42
    500
   1000
   105.00 sq. ft.

Explanation. • Changed to a decimal, 42% becomes .42. • Multiply by 250 and place the decimal point. • The amount to allow for waste is 105 sq. ft.

Illustration. How much is 7% of 745?

Solution.   7% = .07   745
$$\underline{\quad.07}$$
52.15

Explanation. • Change 7% to a decimal (.07) and multiply by 745. • The answer is 52.15.

COMPARING NUMBERS BY PERCENT

There are many cases in which a carpenter uses percent to compare numbers, such as estimating materials and figuring costs.

To compare 2 numbers by percent, place the numbers in fractional form, using the base as the denominator. The base usually follows the word *of*. Divide the numerator by the denominator, and then express the answer as a percent.

Illustration. What percent of 20 is 4?

Solution.   4/20 =   .2 = 20%
$$20\overline{)4.0}$$
$$\underline{4.0}$$

Explanation. • Place the numbers (20 and 4) in fractional form, 4/20. • Divide as usual to get the quotient (.2). • Change the decimal to a percent by moving the decimal point 2 places to the right, and annexing a percent sign.

Illustration.   The total cost of building a house is $19,000. If masonry costs total $4940, what percent of the total cost is used for masonry?

Solution.   4940/19000

$$.26 = 26\%$$
$$19,000\overline{)4940.00}$$
$$\underline{3800\ 0}$$
$$1140\ 00$$
$$\underline{1140\ 00}$$

Explanation. • Divide as usual. • Move the decimal point 2 places to the right, and annex the percent sign.
    • This method may also be used if the numerator is larger than the denominator.

Illustration. What percent of 4 is 20?

Solution.   20/4 = 20 ÷ 4 = 5 = 500%

Explanation. • Divide as above. • Move the decimal point in the quotient two places to the right. • The answer is 500%.

DETERMINING A NUMBER OR QUANTITY WITH A GIVEN PERCENT

To find a number when a certain percentage of that number is given, divide the number by the given percent, expressed as a decimal.

Illustration. 8 is 4% of what number?

Solution. 8 ÷ .04 = 200

Explanation. • Change 4% to a decimal (.04). • Divide as usual. The answer is 200.

APPLICATION

1. Change the following percents to decimals.

   a. 4%   e. 7%   i. 140%   m. 105.7%   q. 1/6%   u. 2/7%
   b. 7%   f. 9%   j. 15.33%   n. 350%   r. 22%   v. 1/13%
   c. 74%   g. 19%   k. 315%   o. 50%   s. 6 1/4%   w. 4 2/9%
   d. 44%   h. 125%   l. 14%   p. 1/3%   t. 6.4 1/3% x. 99%

2. Change the following percents to fractions and reduce to lowest terms.

   a. 7%   e. 5%   i. 10%   m. 18%   q. 80%   u. 108.5%
   b. 8%   f. 13%   j. 12%   n. 19%   r. 90%   v. 31 2/3%
   c. 10%   g. 15%   k. 14%   o. 21%   s. 95%   w. 45 1/2%
   d. 9%   h. 17%   l. 16%   p. 75%   t. 4/7%   x. 225%

3. Change the following fractions to decimals and then to percents.

   a. 1/2   d. 5 3/8 g. 7/8   j. 3 5/7   m. 5/32   p. 16/32
   b. 4/5   e. 5/7   h. 11/16   k. 3/8   n. 5/6   q. 3 3/8
   c. 9/16   f. 1/8   i. 3 5/6   l. 7/9   o. 5/4   r. 1 1/2

4. Answer the following questions concerning percents.

   a. What percent of 75 is 45?   i. 5 is 15% of what number?
   b. What percent of 60 is 45?   j. 10 is 10% of what number?
   c. What percent of 60 is 36?   k. 1 is 15% of what number?
   d. What percent of 32 is 8?   l. 20 is 6% of what number?
   e. What percent of 50 is 24?   m. 9 is 27% of what number?
   f. What percent of 24 is 4?   n. 75 is 50% of what number?
   g. What percent of 16 is 12?   o. 125 is 25% of what number?
   h. What percent of 80 is 80?   p. 155 is 32% of what number?

5. What is 29% of 74″?

6. What is 1% of $895.00?

7. A rafter is 161.04″ long from the ridge to the top plate. If the tail comprises 1% of the length of the rafter, how long is the tail?

8. If a motor on an electrical power saw has 105 horsepower when operated at full speed, what is the horsepower when the machine is operated at 61% of the full speed?

9. The capacity of a dump truck is 4800 lb. If the load it contains is 105% of the capacity, what is the weight of the load?

10. The floor area of a house is 1860 sq. ft. How much total flooring in square feet is needed if 35% more is added for waste?

11. Sheetrock® or fiberboard is used by carpenters as a base for interior siding. If a carpenter receives 27 sheets for a job and 33 1/3% of the sheets cannot be used because of flaws, how many sheets are usable?

12. If a total of 50 sq. ft. of flooring is to be nailed, what percent of flooring is complete after 4 sq. ft. are nailed?

13. The horizontal band saw in figure 21-5 weighs 200 lb. with its legs attached and 150 lb. without its legs. What percent of its weight is comprised by the legs?

Fig. 21-5 Horizontal Band Saw

14. A typical electrical layout for a house is shown in figure 21-6. The contractor agrees to wire the house for $1000. Because of a cash payment, a discount of $100 is given. What percent of the total cost is the discount?

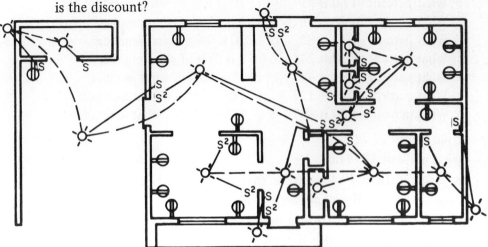

Unit 22
Percent and Money Management

OBJECTIVES

After studying this unit, the student should be able to

- find the interest on a sum of money.
- determine the principal on a loan when the rate of interest and time factor are known.
- find the rate of interest on a loan.

Persons working in the construction industry will probably deal at times with banks and savings and loan institutions. Many principles of banking can be understood by applying information on percentage.

DETERMINING INTEREST ON A LOAN

Interest may be defined as a certain percent of a loan which is paid back to the loaning institution in addition to the loan itself. The percent of interest which is charged is the *rate of interest*. The amount of money which is borrowed, not counting the interest, is known as the *principal*.

To find interest for one year on a sum of money, multiply the amount borrowed by the rate of interest.

Illustration. A carpenter borrows $120 from a bank at the rate of 6% interest per year to purchase tools. How much must be paid in interest per year for the use of the $120?

Solution.   6% = .06 $120 amount borrowed
                            .06 rate of interest
                    $7.20 interest per year

Explanation. • Since the percent sign stands for hundredths, change the 6% to .06 and multiply by $120. • The answer is $7.20.

In some cases, the loan may be paid back in less time than the loaning institution requires. When this happens, determine the interest rate by the month and multiply that figure by the number of months that the loan has been in effect.

Illustration. To buy building equipment, $500 is borrowed from the bank at an interest rate of 7% for one year. The loan is paid back in 3 months. How much is the total interest charge?

Solution.   7% = .07 $500 amount borrowed
                              .07 rate of interest
                      $35.00

$35 \div 12 = 2.916$ or $2.92 monthly interest

$2.92 monthly interest x 3 months = $8.76 total interest charge

FIRST NATIONAL BANK
ANSON, LOUISIANA

LOAN APPLICATION

Date_____ Taken by_____

Name of Applicant_____ Phone_____

Address_____

Age _____ Single _____ Married _____ No. of Dep. _____

Employed by_____ How long_____ Salary $_____
Employer's Address_____
Wife Employed by_____ How long_____ Salary $_____
Employer's Address_____
Other Income _____ $_____
Prev. Employer_____ Total Income . . . . $_____

Wife's Name_____ Race _____
Landlord or Mortgagee_____ Amount of Mortgage $_____
Name and Address of nearest
relative with whom not living_____

LOAN DESCRIPTION

Amount_____
Security_____
Terms_____

REFERENCES

Name _____ Address _____
Name _____ Address _____
Name _____ Address _____
Name of Bank_____

Investigation by Credit Bureau ☐   By 1st National Bank ☐   Date of Report_____

TRADE CLEARANCE

CREDIT WITH	DATE OPENED	HIGH CREDIT	PAYMENTS	PRESENT BALANCE	PAYING RECORD

Approved [ ]   Rejected [ ]   By_____

Fig. 22-1 Typical Loan Application

THE FIRST NATIONAL BANK, ANSON, LA.

PURCHASER'S CREDIT STATEMENT

DEALER _____ DATE _____

NAME IN FULL		AGE	WIFE'S/HUSBAND'S FIRST NAME		DEPENDENTS— AGE	RACE

HOME ADDRESS _____ HOW LONG _____ PHONE _____

FORMER ADDRESS _____ HOW LONG _____ IN CITY HOW LONG _____

NAME AND ADDRESS OF NEAREST RELATIVE NOT LIVING WITH YOU _____

EMPLOYED BY _____ ADDRESS _____ HOW LONG _____

OCCUPATION _____ DRIVERS LICENSE NO. _____ SOCIAL SECURITY NO. _____

BUSINESS PHONE _____ SALARY PER _____ ☐ HOUR ☐ WEEK ☐ MONTH PAY DAYS _____ OTHER INCOME AND SOURCE _____

FORMER EMPLOYER _____ OCCUPATION _____ HOW LONG _____

☐ RENTING ☐ FURNISHED
☐ BUYING ☐ UNFURNISHED
NAME AND ADDRESS OF MORTGAGE HOLDER OR LANDLORD _____ IF BUYING VALUATION _____ AMT. MTG. _____ MO. PMT. OR RENT _____

NAME AND ADDRESS OF BANK WHERE ACCOUNT IS CARRIED _____ ☐ CHECKING ☐ SAVINGS ☐ LOAN YR. AND MAKE CAR _____ LIEN HOLDER _____

WIFE/HUSBAND EMPLOYED BY _____ ADDRESS _____ HOW LONG _____

OCCUPATION _____ CLOCK OR DEPT. NO. _____ NAME AND TITLE OF SUPERIOR _____

BUSINESS PHONE _____ SALARY PER _____ ☐ HOUR ☐ WEEK ☐ MONTH PAY DAYS _____ WIFE'S/HUSBAND'S OTHER INCOME AND SOURCE _____

NAMES OF STORES, BANKS AND FINANCE COMPANIES DEALT WITH	ADDRESS	BALANCE OWING	MO. PMT.	AMOUNT PAST DUE

DEALER'S WORK SHEET

DEALER FILLS IN BEFORE COMPLETING CONTRACT   DATE _____ 19 ___

Description of Merchandise _____ New ☐ Used ☐

CASH SALE DELIVERED PRICE (including Sales Tax, accessories or extras, if any) - - - - - - $ _____ (1)

TOTAL DOWN PAYMENT, $ _____ (NET TRADE-IN) $ _____ (CASH) - - - - - - - - - _____ (2)

DESCRIPTION OF TRADE-IN

Make _____ Model _____ Year _____

DIFFERENCE BETWEEN ITEMS 1 AND 2 - - - - - - - - - - - - - - - - - - - _____ (3)

OFFICIAL FEES - _____ (4)

PRINCIPAL BALANCE (Add Items 3 and 4) - - - - - - - - - - - - - - - _____ (5)

FINANCE CHARGE - _____ (6)

CREDIT LIFE INSURANCE - _____ (7)

AMOUNT OF CONTRACT (Time Balance) - - - - - - - - - - - - - - - - _____ (8)

PAYABLE IN _____ INSTALMENTS OF $ _____

TOTAL TIME PRICE (Add Items 2 and 7) - - - - - - - - - - - - - - - _____ (9)

DEALER SIGNS ☞ _____

BY _____
(IF CORP. OR PART.) (TITLE)

TOM GATES MONROE

Fig. 22-2 Credit Statement

Explanation. • To find the interest for one month, divide the amount of interest for one year by 12 since there are 12 months in a year. Multiply $2.92 by 3 months to find the total paid in interest, $8.76.

Persons in carpentry, especially those who own or operate their own businesses, may need to apply for a loan from a bank or some other loaning institution. Figure 22-1, page 128, shows a typical application for a loan. If equipment is purchased from a store and a loan is necessary, the person taking out the loan may deal with a bank and be required to fill out a credit statement, figure 22-2, page 129.

DETERMINING PRINCIPAL

Sometimes the interest, rate of interest and time are known, but not the prinicpal. The principal is the sum of money being borrowed. To find the principal when the rate of interest and time factor are known, first determine the interest for 1 year by dividing the total interest by the number of years. Then divide the quotient by the given rate of interest, and multiply that quotient by 100.

Illustration. The interest on a 2-year loan at a rate of 7% is $70.00. Find the principal.

Solution. $70 ÷ 2 = $35.00 $35 ÷ 7 = $5.00 $5.00 x 100 = $500

Explanation. • Divide the interest ($70) by 2. • Divide the quotient ($35.00) by 7% which gives the interest for one percent ($5.00). • Finally multiply $5.00 by 100 to determine the principal ($500).

DETERMINING THE RATE OF INTEREST

To find the interest rate on a loan, divide the interest by the principal for 1 year and change the decimal to a percent.

Illustration. A carpenter borrows $8000 for one year and pays $360 interest. What is the rate of interest?

Solution. $360/$8000 = 360/8000 = 9/200

$$\begin{array}{r} .04\ 100/200 \\ 200\ \overline{)9.00} \\ \underline{8\ 00} \\ 1\ 00 \end{array}$$

.04 100/200 = .04 1/2 = 4 1/2%

Explanation. • The principal ($8000) becomes the denominator of the fraction, and the interest ($360) is the numerator. • The numerator divided by the denominator equals .04 1/2. • Change the quotient to a percent. • The answer is 4 1/2%.

DETERMINING THE TIME FACTOR

It is possible to determine the time that a loan is in effect when the principal, interest, and interest rate are known.

Illustration. In what amount of time will $500 gain $100 interest at 8%?

Solution. $500 principal $100 total interest   2.5
                 .08 rate $40 interest for one year   40)100.0
              $40.00 interest for one year   80
                                                                      20 0
              Time = 2.5 = 2 1/2 years   20 0

Explanation. • Change 8% to .08 and multiply by the principal ($500.00). • The product ($40.00) is the interest for one year. • Divide the total interest ($100.00) by the interest for one year. • The quotient is 2 1/2, the time in which $500.00 would gain $100.00 in interest at a rate of 8%.

DISCOUNTS

When a tradesman frequently buys materials from a certain place and pays within a certain time, a discount is sometimes given. A discount is a sum of money subtracted from the original price, usually figured on a percent rate.

To find the amount of a discount, multiply the regular price by the percent of discount.

Illustration. Determine a 3% discount on $350 spent for building materials.

Solution.   3% = .03 $350 x .03 = $10.50

Explanation. • Change 3% to .03, and multiply the total amount ($350) by the rate of interest. • The amount of discount is $10.50.

• Subtract the discount from the regular price to determine the net, or selling, price.

APPLICATION

1. A group of carpenters begin construction of a frame, figure 22-3. After one week of construction, 3500 bd. ft. of lumber are used, which is 25% of the total amount for the project. How many board feet of lumber are required for the entire project?

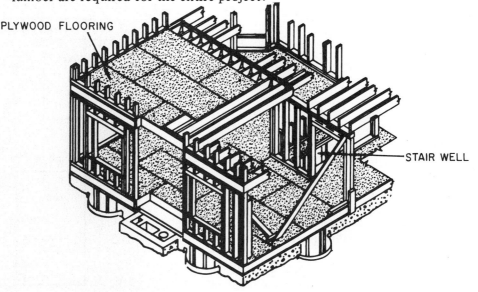

PLYWOOD FLOORING

STAIR WELL

Fig. 22-3

2. If a contractor borrows $60.00 at 6%, how much interest is paid in one year?

3. A house is built for $12,000 and sells for a 20% profit. What is the profit?

4. How much interest is paid on $635 in 4 months at a rate of 6.5% per year?

5. How much interest is paid on $635 in 18 months at a rate of 6.5% per year?

6. A carpenter borrows money from a bank at 6% interest. At the end of 2 years, the amount of interest paid is $216. What is the principal?

7. What is the rate of interest on a $600, one-year loan if $120 in interest is paid in 6 months?

8. In what amount of time will $900 gain $144 interest at a rate of 8%?

9. In what amount of time will $900 gain $36 interest at a rate of 8%?

10. A building supply company sells a carpenter $95.00 worth of siding at a 4% discount. How much is the discount?

11. A carpenter purchases 5000 bricks. The selling price is $60 per 1000. At a 3% discount, what is the net price?

12. A kit of carpentry tools that costs $120 is sold at a discount of $12. What is the percent of discount?

13. Kitchen cabinets must be built with convenience and appearance in mind, figure 22-4. They may arrive on the job site prefabricated, which the carpenter installs or be constructed by the carpenter himself. A carpenter purchases materials for a cabinet worth $200.00. He is given a 5% discount. What does he pay for the supplies?

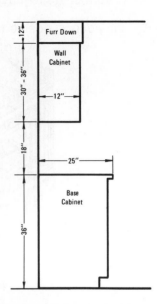

Fig. 22-4 Typical Kitchen Cabinet.
Most kitchen cabinets
are built to the same basic
dimensions.

CAREER PROFILE: THE TRIMMER

Job Description

A trimmer may specialize in interior trim or exterior trim or work in both areas. Interior trimming includes the installation of door jambs and trimming, molding, and paneling. The interior trimmer may apply finish to walls and ceilings. Many times, they must be responsible for calculating materials needed for a job.

Exterior trimmers install window frames, exterior door frames, and siding. They may construct cornices and corner boards. On a large project, trimmers may have a very specialized job, such as installing exterior siding.

Qualifications

Besides general training in carpentry, trimmers should possess specialized knowledge in such areas as construction of wood joints, tools, and materials. A good sense of balance and manual dexterity are essential, since much of the work of the trimmer is done on scaffolds and ladders.

Interior Trim—Trimming out the Door Casing on the inside of the Home

Section 4 The English System of Measurement

Unit 23   Using the English System: Linear, Volume, and Weight Measurements

OBJECTIVES

After studying this unit, the student should be able to

- change linear units of one denomination to linear units of another denomination.
- change volume measurements of one denomination to volume measurements of another denomination.
- add, subtract, multiply, and divide linear, volume, and weight measurements of different denominations.

UNITS OF MEASUREMENT: LINEAR

A unit of measurement is a standard by which length, area, or volume is measured. Linear measurement refers to those objects which have only the dimension of length. Materials such as rods, moldings, and wire are sold by length. Materials which are less than 2″ wide are sold by length.

Common standards of linear measurement used in carpentry include the following:

$$12 \text{ inches (in. or ″)} = 1 \text{ foot (ft. or ′)}$$
$$36 \text{ inches} = 1 \text{ yard (yd.)}$$
$$3 \text{ feet} = 1 \text{ yard}$$

CONVERTING UNITS TO DIFFERENT DENOMINATIONS

Since only units of the same denomination may be added, subtracted, multiplied, or divided, it is necessary for the carpenter to know how to express the same quantity in different terms. For example, 12″ may be expressed as 12″ or 1 foot.

To change a known linear unit measurement of one denomination to a unit of another denomination, multiply the number of units by the proper conversion value.

Illustration. Shown in figure 23-1 is a closet door 2′ wide. Change this measurement to inches.

Solution. $1′ = 12″$
$2′ = 2 \times 12″ = 24″$

Explanation. • Convert feet to inches and multiply by the number of units (2). • The answer is 24″.

The same basic procedure may be used to convert different denominations, such as feet to yards or yards to inches.

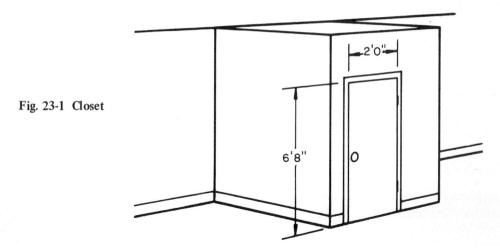

Fig. 23-1 Closet

Illustration. A bathroom cabinet is 2′ 6″ long. Change this measurement to inches.

Solution.   2 x 12 + 6 = 30″

Explanation. • Follow the procedure for changing feet to inches by multiplying 2′ by 12 and adding 6″. • The answer is 30″. • The phrase 12″ = 1′ is known as a *conversion factor.*

Illustration. A standard inside door is 6′ 8″ in height. Change this measurement to feet.

Solution.   6′ + 8/12′ = 6 2/3′

Explanation. • Follow the procedure for changing inches to feet by dividing the number of inches given by 12. • In this case, the terms are expressed as a common fraction and reduced to lowest terms (8/12 = 2/3). • Annexing the 2/3′ to 6′ gives the answer, 6 2/3′.

Illustration. Change 5 yd. 2′ to feet.

Solution.   5 x 3 + 2 = 17′

Explanation. • To change yards to feet, multiply the number of yards given (5) by 3 and add the 2′.

Illustration. The pipe diagram in figure 23-2 is drawn to a scale of 1/4″ = 1′ 0″. If the length of the main line is 30′, how many linear yards are there?

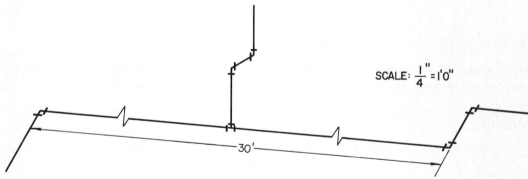

SCALE: $\frac{1}{4}'' = 1'0''$

Fig. 23-2 Pipe Diagram

Solution. 3' = 1 yd. 30' ÷ 3' = 10 yd.

Explanation. • To change linear feet to yards, apply the conversion factor and divide the number of feet by 3.

Illustration. Tri-level construction, used in residential construction, requires a combination of different types of framing because of shrinkage and settling. The distance from floor to ceiling in figure 23-3 represents 10 yd. Change this measurement to inches.

Solution. 36" = 1 yd. 10 x 36" = 360"

Explanation. • In this case, 36" = 1 yd. is considered the conversion factor. • Multiply 36 by the number of yards (10).

Fig. 23-3 Tri-level Construction

Illustration. The length of an electrical cable extending from street to house is 3600" long. How many linear yards are there in this measurement?

Solution. 36" = 1 yd. 3600 ÷ 36 = 100 yd.

Explanation. • To change linear inches to linear yards, divide the number of inches by 36.

FUNDAMENTAL PROCESSES IN LINEAR MEASUREMENT

There are certain recommended procedures to use when solving problems in linear measurement requiring addition, subtraction, multiplication, and division.

Addition

Illustration. A *conduit pipe* is a channel for carrying fluids. It is also used to protect electrical wires. Sections of the conduit pipe and fitting in figure 23-4 measure 10' 5" and 6' 9". What is the total length?

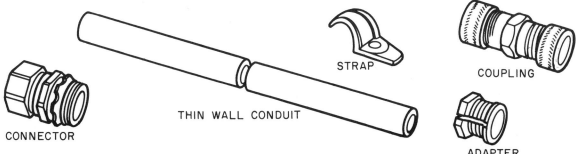

STRAP

COUPLING

THIN WALL CONDUIT

CONNECTOR

ADAPTER

Fig. 23-4 Conduit Pipe and Fittings

Solution. Step 1. 10′ 5″ + 6′ 9″ = 16′ 14″
Step 2. 14″ ÷ 12 = 1′ 2″
Step 3. 16′ + 1′ 2″ = 17′ 2″

Explanation. • Add the 2 measurements (10′ 5″ and 6′ 9″). • Since the inches total more than 1′, divide by 12. Add this to the total number of feet (16) for the full measurement of the pipe.

Subtraction

Illustration. Subtract 2 yd. 2′ from 5 yd. 1′.

Solution. 5 yd. 1′ = 4 yd. 4′
2 yd. 2′ = 2 yd. 2′
2 yd. 2′

Explanation. • Since the 2′ in the subtrahend is larger than 1′ in the minuend, borrow 3′ from 5 yd. which leaves 4 yd. Add 3′ and 1′, totaling 4′. • Then subtract 4 yd. 4′ from 2 yd. 2′. • The answer is 2 yd. 2′.

Multiplication

There are three procedures from which to select when multiplying numbers expressed in two or more units of measurement. One procedure is to multiply each measurement separately and then convert the smaller unit to the larger. Another is to express both units as a mixed number. The other procedure is to multiply after changing all units to the smaller denomination.

Illustration. If an object is comprised of 4 equal pieces measuring 1′ 8″ each, what is the length of the whole object?

Solution. Step 1. 1′ 8″ x 4 = 4′ 32″
Step 2. 32″ ÷ 12 = 2′ 8″
Step 3. 4′ + 2′ 8″ = 6′ 8″

Explanation. • Multiply 1′ 8″ by 4: 8″ x 4 = 32″; 1′ x 4 = 4′. • Change 32″ to 2′ 8″ and add 4′. • The total length is 6′ 8″.

Illustration. Multiply 1′ 8″ by 4.

Solution. Step 1. 1′ 8″ = 1 2/3′
Step 2. 1 2/3′ x 4 = 5/3 x 4 = 20/3 = 6 2/3′

Explanation. • Before multiplying 1′ 8″ by 4, change 1′ 8″ to 1 2/3′. The product is 20/3′, which equals 6 2/3′.

Illustration. Multiply 1′ 8″ by 4.

Solution. (1′ = 12″)
Step 1. 12″ + 8″ = 20″
Step 2. 20 x 4 = 80″
Step 3. 80″ ÷ 12 = 6′ 8″ = 6 2/3′

Explanation. • Change 1′ to 12″ and add 8, giving 20″. • Multiply 20 by 4, which equals 80″ and divide by 12. The answer is 6′ 8″ or 6 2/3′.

Division

To divide two units of different denominations, change the units to the smaller denomination and divide by the given number.

Illustration. *Rafters* are roof members which support the covering of a roof, figure 23-5. The total rafter length shown in the illustration, measured from point A to point B, is 7′ 6″. If the rafter is divided into 4 equal parts, what is the measurement of each part?

Solution. Step 1. 7′ x 12 = 84″
Step 2. 84″ + 6″ = 90″
Step 3. 90″ ÷ 4 = 22 1/2″

Explanation. • Change 7′ to 84″ and add 6″, equaling 90″. • Divide 90″ by 4, giving the answer of 22 1/2″.

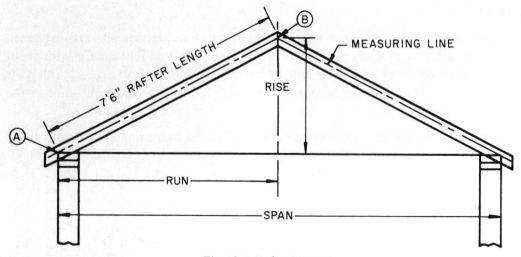

Fig. 23-5 Rafter Measurement

UNITS OF MEASUREMENT: VOLUME

The major units of volume in the English system of measurement used in construction are pints, quarts, and gallons. To change units of one denomination to units of another denomination, follow the basic procedure shown in the section on linear units. Consult the table and multiply or divide by the proper number. Examples are shown in changing gallons to quarts and quarts to gallons, and pints to quarts and quarts to pints. Figure 23-6 shows these equivalents and others in the English system of measurement.

2 pints = 1 quart (qt.)
4 quarts = 1 gallon (gal.)
31 1/2 gallons = 1 barrel (bbl.)
1 gallon = 231 cubic inches
7 1/2 gallons = 1 cubic foot
1 gallon of water = 8 1/3 lb.
1 cu. ft. of water = 62 1/2 lb.

Fig. 23-6 Volume Measurement Equivalents – English System

Changing Gallons to Quarts and Quarts to Gallons

To change gallons to quarts, multiply the number of gallons by 4.

Illustration. Concrete is hardened by a chemical reaction between sand, gravel, and cement. There are 2 gal. of a concrete mixture of which 7 qt. are water. What part of the mixture is not water? Express the answer in quarts.

Solution.   Step 1. 2 x 4 qt. = 8 qt.
                        Step 2. 8 qt. – 7 qt. = 1 qt.

Explanation. • There are 4 quarts in one gallon. Multiply 4 qt. by 2 gal. to find the number of quarts. • Then subtract 7 qt. from 8 qt. which leaves 1 qt.

To change gallons to quarts, divide the number of quarts by 4.

Illustration. Change 12 qt. to gallons.

Solution.   12 ÷ 4 gal. = 3 gal.

Explanation. • There are 4 qt. in 1 gal., so divide 12 by 4. • The result is 3 gal.

Changing Pints to Quarts and Quarts to Pints

To change pints to quarts, divide the number of pints by 2.

Illustration. Change 6 pt. to quarts.

Solution.   6 ÷ 2 = 3 qt.

Explanation. Divide the number of pints (6) by the number of pints in 1 qt. (2) for an answer of 3 qt.

To change quarts to pints, multiply the number of quarts by 2.

Illustration. Change 7 qt. to pints.

Solution.   7 x 2 = 14 pints

Explanation. Multiply the number of quarts (7) by 2. The number of pints in 7 qt. is 14.

DETERMINING VOLUME EQUIVALENTS

The volume of a container can be determined by multiplication or division.

Volume of Gallons, Cubic Inches and Cubic Feet

To determine the number of gallons equal to a certain number of cubic inches, divide the number of cubic inches by 231.

Illustration. The volume of a container of water is 924 cu. in. Find its volume in gallons.

Solution.   924 ÷ 231 = 4 gal.   *Check:* 231 x 4 = 924 cu. in.

Explanation. • Since 1 gal. is equivalent to 231 cu. in., divide 924 gal. by 231. • The answer is 4 gal.

To determine the number of cubic inches in a gallon, multiply the gallon by 231.

Illustration. How many cubic inches are equal to 5 gal.?

Solution.   231 cu. in. x 5 = 1155 cu. in.

Explanation. • Multiply 231 cu. in. by the given number of gallons (5). • The product is 1,155 cu. in.

The equivalent of 7 1/2 or 7.5 gal. of water is 1 cu. ft. To determine the volume in gallons equal to a given volume in cubic feet, multiply the given number of cubic feet by 7.5.

Illustration. How many gallons are contained in a tub which holds 12 cu. ft. of water?

Solution.   12 x 7.5 gal. = 90 gal.

Explanation. • Multiply the number of cubic feet (12) by the number of gallons in 1 cu. ft. (7.5). • The product is 90 gal.

To find the volume in cubic feet which is equal to a given number of gallons, divide the number of gallons by 7.5.

Illustration. Find the number of cubic feet of water in 90 gal.

Solution.   90 ÷ 7.5 = 12 cu. ft.

Explanation. • Divide 90 gal. by 7.5 gal. • The answer is 12 cu. ft.

UNITS OF MEASUREMENT: WEIGHT

The weight of an object is produced by a straight downward pull called *gravity*. In carpentry, weight is measured on various types of scales. For example, scales with level arms are used to weigh heavy objects, and spring-balanced scales are used for light objects.

There are two systems of weight measurement used in the construction trade: the *avoirdupois,* which is the English system, and the metric system. In the English system, common units of measurement include the ounce, pound, and ton. Figure 23-7 gives basic measurements of weight.

16 ounces = 1 pound (lb.)
2000 pounds = 1 short ton (s.t.)
2240 pounds = 1 long ton (l.t.)

Fig. 23-7

To change measurements, multiply or divide by the proper number.

Changing Pounds to Ounces

Pounds and ounces are the most frequently used units of weight measurement in construction. Some tools are labelled in ounces, such as the 13 oz. hammer.

To change pounds to ounces, multiply the number of given pounds by 16.

Illustration. Change 5 lb. to ounces.

Solution. 5 x 16 oz. = 80 oz.

Explanation. • Multiply 5 by 16 oz. since there are 16 oz. in 1 pound. • The answer is 80 oz.

Changing Ounces to Pounds

Ounces are changed to pounds by dividing the number of ounces by 16.

Illustration. An object on a set of scales weighs 36 ounces. How many pounds does it weigh?

Solution. 36 ÷ 16 = 2 1/4 lb.

Explanation. • Since one pound is equivalent to 16 oz., change 36 oz. to pounds by dividing 36 oz. by 16. The quotient is 2 lb. 4 oz. • The 4 oz. is changed to 4/16, which equals 1/4 lb. • The answer is 2 1/4 lb.

Changing Short and Long Tons to Pounds and Pounds to Short and Long Tons

The pound may be used to weigh supplies used in construction, such as nails, lead, copper, and roofing materials. The ton is used to weigh objects such as concrete piles and steel beams.

There are two types of tons: the *short* ton, which has a weight of 2000 lb., and the *long* ton, which has a weight of 2240 lb. To change short tons to pounds, multiply the number of short tons by 2000.

Illustration. A steel *truss* is a combination of steel members which supports loads over the span of a roof, figure 23-8. A certain steel truss weighs 1 1/2 tons. How many pounds does it weigh?

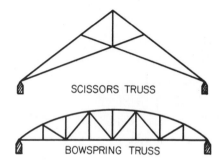

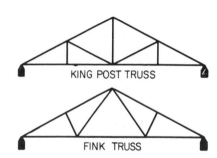

Fig. 23-8 Various Steel Trusses

Solution.   1 1/2 x 2000 = 3/2 x 2000 = 3000 lb.

Explanation. • Change 1 1/2 s.t. to 3/2 s.t. and multiply by 2000. • The answer is 3000 lb.

To change short tons to pounds, divide the number of pounds by 2000.

Illustration.   A tag on a concrete pile is marked 3500 lb. Find the number of short tons in the pile.

Solution.   3500 ÷ 2000 = 1.75 s.t. or 1 3/4 s.t.

Explanation. • Divide 3500 lb. by 2000 lb. • The result is 1 3/4 s.t.

To change long tons to pounds, multiply the number of long tons by 2240.

Illustration. A truck loaded with 5 cu. yd. of concrete weighs 2 l.t. How many pounds are there in the truck?

Solution.   2 x 2240 lb. = 4480 lb.

Explanation. • There are 2240 lb. in 1 l.t. • Multiply the number of long tons (2) by 2240, which gives an answer of 4480 lb.

To change pounds to long tons, divide the number of pounds by 2240.

Illustration. Change 2800 lb. to long tons.

Solution.   2800 ÷ 2240 = 1 1/4 l.t.

Explanation. • To change 2800 lb. to long tons, divide 2800 by 2240. • The result is 1 ton and 560 lb., or 1 1/4 l.t.

To add, subtract, multiply, and divide volume or weight measurements, follow the procedure shown in the section on linear units. Consult the table until units of measurement are memorized.

APPLICATION

1. Change each measurement to linear inches.

    a. 4' d. 12' g. 3/4' j. 5/6' m. 8'
    b. 9' e. 13' h. 3/8' k. 6' n. 2'
    c. 10' f. 1/2' i. 5/8' l. 7' o. 11'

2. Change each measurement to linear feet.

    a. 36" d. 28" g. 10" i. 72" k. 54"
    b. 40" e. 120" h. 11" j. 60" l. 84"
    c. 42" f. 166"

3. Solve each problem involving linear measurements.

    a. 2 ft. from 1 yd.   h. 100 yds. ÷ 1/3
    b. 5 yds. 1' from 9 yd.   i. 5' 9" + 6' 11" + 8' 5"
    c. 23 yds. 2' from 33 yds.   j. 9' 9" + 8' 7" + 4' 11"
    d. 28 yds. ÷ 7 yds.   k. 6 yds. 2' + 5 yds. 1' + 7 yds. 2'
    e. 12 yds. 2' x 3   l. 12' 3" – 8' 9"
    f. 15 yds. 2/3' x 3/4   m. 9' 1" – 6' 3"
    g. 23 yds. ÷ 4   n. 21' 2" – 14' 5"

4. Change each of the following volume measurements.

   a. 6 pts. to quarts   e. 52 pints to gallons
   b. 3 qt. to pints   f. 462 cu. in. to gallons
   c. 17 gal. to pints   g. 16 quarts to gallons
   d. 6 gal. to cubic inches

5. Change the following measurements of weight.

   a. 6 lb. to ounces
   b. 8 lb. to ounces
   c. 29 lb. to ounces
   d. 32.5 lb. to ounces
   e. 48 oz. to pounds
   f. 32 oz. to pounds
   g. 56 oz. to pounds
   h. 60 oz. to pounds
   i. 8 s.t. to pounds
   j. 9 s.t. to pounds
   k. 13 s.t. to pounds
   l. 24 s.t. to pounds
   m. 12,000 lb. to short tons
   n. 35,000 lb. to short tons
   o. 40,000 lb. to short tons
   p. 4 l.t. to pounds
   q. 5 l.t. to pounds
   r. 7 l.t. to pounds

Unit 24
Perimeter

OBJECTIVES

After studying this unit, the student should be able to

- find the perimeter of squares, rectangles, triangles, and circles.

Perimeter may be defined as the distance around an object. *Formulas,* sets of numbers which briefly express a concept, may be used to figure the perimeter of squares, rectangles, triangles, and circles.

FINDING THE PERIMETER OF A SQUARE

A *square* is a rectangle which has four equal sides and four 90° corners. The formula for finding the perimeter of a square is $P = 4S$.

$$P = \text{perimeter}$$
$$S = \text{side}$$

Illustration. The drawing in figure 24-1 has 4 equal sides. What is the perimeter if each side measures 5'?

Solution. $P = 4S$
  Step 1. $S = 5'$
  Step 2. $P = 4 \times 5 = 20'$

Explanation. • To find the perimeter (P), multiply the number of sides (4) by the length of each side (5'). • The perimeter of the square is 20'.

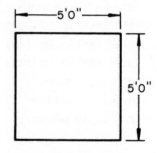

Fig. 24-1 Square

FINDING THE PERIMETER OF A RECTANGLE

A rectangle is used in carpentry problems as frequently as the square. A *rectangle* is an object or figure with two sides equal in length and two sides equal in width, and four 90° angles. The formula for finding the perimeter of a rectangle is $P = 2l + 2w$.

Illustration. The length of the layout in figure 24-2 is 20' and the width is 10'. What is the perimeter in feet?

Solution.   Step 1. $P = 2l + 2w$
  Step 2. $P = 2 \times 20 + 2 \times 10$
  Step 3. $P = 40 + 20 = 60'$

Explanation. • The perimeter is found by multiplying the length (20') twice, multiplying the width (10') twice, and adding the products. • The answer is 60'.

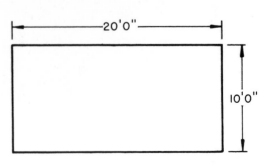

Fig. 24-2 Rectangle

144

FINDING THE PERIMETER OF A TRIANGLE

A *triangle* is a three-sided figure with three angles equaling 180°. Roofs of some houses and various parts of the roof are examples of triangular-shaped objects. The formula for finding the perimeter of a triangle is P = a + b + c, which represent the lengths of each side.

Illustration. Find the perimeter of the triangle in figure 24-3.

Solution. P = 3′ + 4′ + 5′ = 12′

Explanation. • In accordance with the given formula, add the three sides. • The perimeter is 12′.

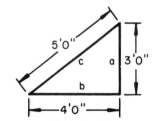

Fig. 24-3 Triangle

An equilateral triangle is a triangle with three equal sides, figure 24-4. The perimeter can be found by the formula P = 3S.

Illustration. Find the perimeter of the equilateral triangle in figure 24-4.

Solution. P = 3 x 4′ = 12′

Explanation. • Multiply the number of sides (3) by the length (4′); the perimeter is 12′.

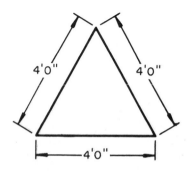

Fig. 24-4 Equilateral Triangle

FINDING THE PERIMETER OF A CIRCLE

The perimeter of a circle is known as the *circumference.* The formula used to find the perimeter of a circle is C = π d or C = 2π r, figure 24-5. The *diameter* (d), which is the distance across the center of a circle, is two times the length of the radius (r). The *radius* is defined as the distance from the center of a circle to surface and is equal to one-half the diameter. The circumference is equal to pi times the diameter. Pi, or π, is equal to 3 1/7 or 3.1416.

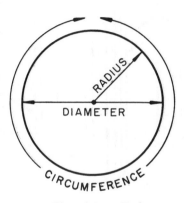

Fig. 24-5 A Circle

Illustration. The diameter of the circle in figure 24-6 is 5″. What is its perimeter?

Solution.   Step 1.   C = 3.1416 x 5
                  Step 2.   3.1416 x 5 =
                                    15.7080″

Explanation. • To find the perimeter of the circle multiply π (3.1416) by the diameter (5″). The product is 15.7080″.

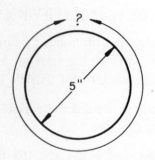

Fig. 24-6 Circle with 5″ Diameter

USING THE PERIMETER FORMULAS IN MATERIAL ESTIMATING

Perimeter formulas can be used to estimate materials in construction, although a slight alteration in the formulas is sometimes necessary.

Plates are horizontal members placed on top of studs on wall frames. To determine the number of lineal feet of 2 x 4s needed for the top plates in figure 24-7, use the formula P = 2 l + 2 w. After the number of plates is determined, multiply by the number of plates. Allow 10% for waste.

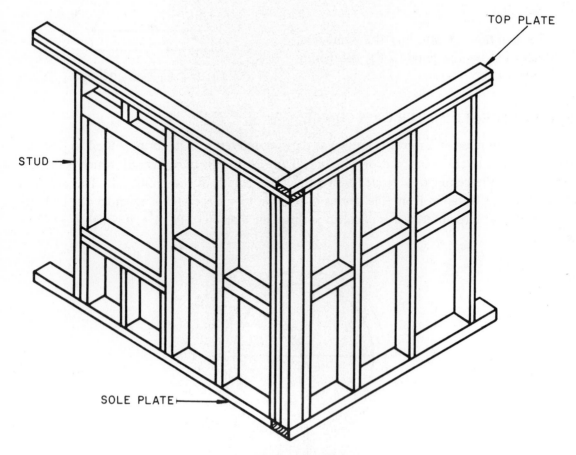

Fig. 24-7 Plates on Rough Framing

Illustration. Find the number of feet of lumber needed for top plates in the building illustrated in figure 24-8.

Solution. Step 1. P = (2 x 14') + (2 x 12') = 52'
Step 2. 52' x 2 = 104'
Step 3. 104' x .10 = 10.4'
Step 4. 104' + 10' = 114'

Explanation. • Two sides of the room measure 14' and two measure 12'; therefore, there is a total of 52 linear feet around the frame. • Since there is a total of two plates, multiply by 2. • To allow for waste, change 10% to .10 and multiply by 104' for total waste, 10.4'. • Disregard the .4 since it is less than half a foot. • Add 104' and 10' for a total of 114'.

Perimeter formulas can be used to estimate the number of studs required for walls. To figure the number of studs needed first apply the formula P = 2 l + 2 w. Then multiply the total number of linear feet by 3/4 when studs are placed 16" O.C. (since 12/16" = 3/4). Add two studs for each corner, door, partition, and window opening.

Illustration. A layout of a rectangular room 14' long and 12' wide is shown in figure 24-8. The structure has one partitioned wall the entire width, two doors, and two windows. Determine the number of studs needed, assuming they are to be placed 16" O.C.

Solution. Step 1. P = 2 l + 2 w = (2 x 14) + (2 x 12) = 52'
Step 2. 52' (3/4) = 39
Step 3. 39 + 8 + 4 + 4 + 2 = 57 studs

Explanation. • Since two walls measure 14' and two measure 12', the perimeter of the room is 52'. • The studs are placed 16" on center; 52 x 3/4 = 39. • There are an additional 8 corner studs, 4 for doors, 4 for windows, and 2 for the partition, totaling 18 studs. • Add the 39 studs and the 18 studs, which gives a total of 57 studs.

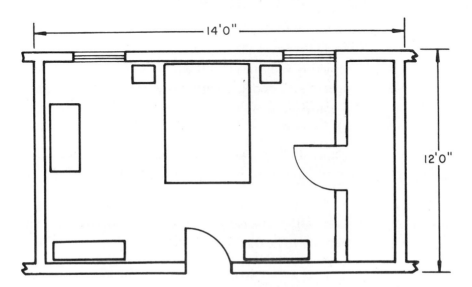

Fig. 24-8 Room Layout

APPLICATION

1. Find the perimeter of rectangles having the following dimensions.

    a. 20" long; 15" wide
    b. 25" long; 14.6" wide
    c. 53' 6" long; 43' 8" wide
    d. 30 yds. long; 12 yds. wide
    e. 17 3/8" long; 12 5/6" wide
    f. 3 2/3 yds. long; 2 7/8" wide
    g. 5 1/2' long; 3 3/4' wide
    h. 5 yds. 2' long; 3 yds. wide
    i. 6 yds. 1' long; 3 yds. 2' wide

2. What is the perimeter of squares which have the following dimensions?

    a. 35'
    b. .45'
    c. 25"
    d. 3.125'
    e. 3.5 yds.
    f. 32 1/2'
    g. 8.9 yds.
    h. 34 1/2" inch
    i. 10 3/4 yds.
    j. 55.5"

3. What is the perimeter of circles which have the following dimensions?

    a. radius 6'
    b. diameter 6'
    c. radius 6"
    d. diameter 12"
    e. radius 6 yds.
    f. diameter 18'
    g. radius 2' 6"
    h. diameter 5'

4. What is the perimeter of a square with sides 14" in length?

5. What is the perimeter of a square with sides 8 3/4' in length?

6. If the length of the sides of an equilateral triangle measure 9", what is the perimeter in inches?

7. If the sides of a square measure 11", what is the perimeter in feet and inches?

8. If the sides of a triangle measure 6", 8", and 10", what is the perimeter?

9. If the diameter of a circle is 9", what is the circumference? (The radius is half the length of the diameter.)

10. The radius of a circle is 5". What is the circumference?

11. What is the perimeter of a rectangle if its length measures 16' and width measures 13' 6"?

12. The layout of a structure indicates that it will be 42' long and 24' wide with 6 window openings, 6 doors, 1 partition extending the length of the layout, and 4 partitioned walls extending the width. How many studs are required if they are to be spaced 16" O.C.?

13. A layout shows that a building will be 30' square. If 3 plates are required, how many linear feet of 2 x 4s are needed? Allow 10% for waste.

14. How many common rafters are needed for a gable roof 30' wide and 46' long if the rafters are to be spaced 24" O.C.?

15. A bench 38' 8" long is to be divided into 18 equal spaces for seats. How much space is allowed for each seat?

16. A carpenter cuts a piece of lumber that is 7′ 9″ long into 7 equal parts. What is the length in feet and inches of each piece? Do not allow for waste.

17. *Weather stripping* is a material placed around doors and windows to prevent drafts. The size of a window opening is 30″ by 60″. How many lineal feet of weather stripping are needed to extend across the top and down each side of the window?

18. A *hip roof,* figure 24-9, is formed by inclining planes on all sides of the structure. The hip on the roof in the figure forms an equilateral triangle. With the given measurement, what is the perimeter of the triangle?

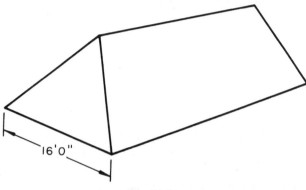

16′0″

Fig. 24-9

19. If the diameter of a disc is 16″, what is the length of wire needed to extend around the disc?

Unit 25   Using the English System:
Square Measurement

OBJECTIVES

After studying this unit, the student should be able to

- change square units of one denomination to square units of another denomination.
- add, subtract, multiply, and divide square units.
- estimate materials by square measurement.

Persons in carpentry usually encounter situations where they must work with square units of measurement. For example, materials such as flooring, roofing, and wall sheathing are sold by the square unit. Generally, square units are used to refer to surface areas.

Think of square units as being areas with two dimensions, length and width. The thickness of the object is not considered in square measurements. For example, the area of a floor has length and width. Equivalents of square unit measurement used in carpentry include the following:

144 square inches (sq. in.) = 1 square foot (sq. ft.)
9 square feet = 1 square yard (sq. yd.)
100 square feet = 1 square

CONVERTING SQUARE FEET TO SQUARE INCHES

There are 144 sq. in. in 1 sq. ft. To change square feet to square inches, multiply the number of square feet by 144.

Illustration. The layout in figure 25-1 is 1' long and 1' wide, or 1 sq. ft. Change 1 sq. ft. to square inches.

Solution.   144 sq. in. x 1 = 144 sq. in.

Explanation. • Count the squares in figure 25-1 by multiplying 12 x 12. • If the square is 1 ft. by 1 ft. or 1 sq. ft., multiply 1 by the conversion factor, 144 sq. in. • The answer is 144 sq. in.

CONVERTING SQUARE INCHES TO SQUARE FEET

To change square inches to square feet, divide the number of square inches by 144.

Illustration. The layout in figure 25-2 has 576 sq. in. Change 576 sq. in to square feet.

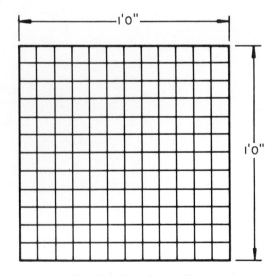

Fig. 25-1 One Square Foot

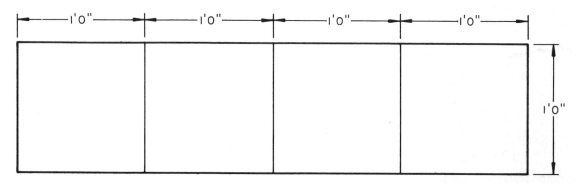

Fig. 25-2 Four Square Feet

Solution. $576 \div 144 = 4$ sq. ft.

Explanation. • There are 144 sq. in. in 1 square foot. • Divide the number of square inches (576) by 144. • The answer is 4 sq. ft.

CONVERTING SQUARE YARDS TO SQUARE FEET

There are 9 sq. ft. in 1 sq. yd. To change square yards to square feet, multiply the number of square yards by 9.

Illustration. Material required to construct a window awning totals 5 sq. yd. How many square feet is this?

Solution. 5×9 sq. ft. = 45 sq. ft.

Explanation. • There are 5 sq. yd. of material. Multiply 5 by 9 which equals 45 sq. ft. • In this case, the conversion factor is 9 multiplied by the given measurement.

CONVERTING SQUARE FEET TO SQUARE YARDS

To change square feet to square yards, divide the number of square feet by 9 since there are 9 sq. ft. per square yard.

Illustration. Change 45 sq. ft. to square yards.

Solution. $45 \div 9 = 5$ sq. yd.

Explanation. • The material is expressed in square feet. Divide the number of feet (45) by 9 sq. ft. to obtain the quotient (5 sq. yd.).

CONVERTING SQUARE YARDS TO SQUARE INCHES

To change square yards to square inches, first change the square yards to square feet by multiplying the number of square yards by 9. Change the result to square inches by multiplying the number of square feet by 144.

Illustration. Express 5 sq. yd. in square inches.

Solution. $5 \times 9 = 45$ sq. ft. $45 \times 144 = 6480$ sq. in.

Explanation. • First, obtain the total number of square feet (45). • Then, convert this figure to square inches by multiplying by 144. • The answer is 6480 sq. in.

CONVERTING SQUARE INCHES TO SQUARE YARDS

When changing square inches to square yards, first convert square inches to square feet by dividing the number of square inches by 144. Then divide the number of square feet by 9.

Illustration. Change 6480 sq. in. to square yards.

Solution.   Step 1. 6480 ÷ 144 = 45 sq. ft.
                Step 2. 45 ÷ 9 = 5 sq. yd.

Explanation. • To change 6480 sq. in. to square yards, first change square inches to square feet. • Divide 6480 by 144, which gives 45 sq. ft. Then convert 45 sq. ft. to square yards by dividing 45 by 9. The answer is 5 sq. yd.

CONVERTING SQUARE FEET TO SQUARES

A *square* is an area equivalent to 100 sq. ft. A square does not have to be in a square shape; squares are found within triangular-shaped objects, rectangles, and circles.

To change square feet to a square, divide the number of square feet by 100.

Illustration. A *butterfly* roof is one which appears to be two connected shed roofs constructed at an angle. There are 500 sq. ft. contained in the butterfly roof shown in figure 25-3. How many squares are in this area?

Solution.   (Squares = square feet/100)

                500/100 = 5 squares

Explanation. • Since there are 100 sq. ft. in one square, 100 is the figure used in the conversion process. Divide 500 by 100 which equals 5 squares.

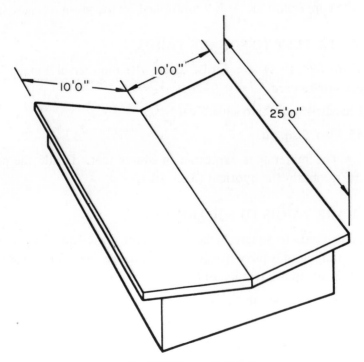

Fig. 25-3 Butterfly Roof

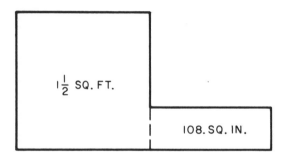

Fig. 25-4 Square Measures

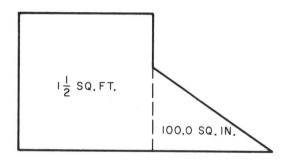

Fig. 25-5 Square Measures

FUNDAMENTAL PROCESSES IN SQUARE UNIT MEASUREMENT

Working with square units involves the processes of addition, subtraction, multiplication, and division.

Addition

Illustration. There are 108 sq. in. in the smaller section of figure 25-4 and 1 1/2 sq. ft. in the larger section. How many total sq. ft. are there in the drawing?

Solution.   Step 1. 108 sq. in. = 108/144 = 3/4 sq. ft.
Step 2. 3/4 sq. ft. + 1 1/2 sq. ft. = 3/4 sq. ft. + 1 2/4 sq. ft. = 1 5/4 sq. ft.
Step 3. 1 5/4 sq. ft. = 2 1/4 sq. ft. = 2 sq. ft. 36 sq. in.

Explanation. • Change 108 sq. in. to a fraction of a foot (3/4 sq. ft.). • Use 4 as the L.C.D. in the addition. Add 3/4 and 1 2/4, the sum of which is 2 1/4. • The measurement may be expressed as 2 1/4 sq. ft. or 2 sq. ft. 36 sq. in.

Subtraction

Illustration. One portion of the drawing in figure 25-5 is 100 sq. in. and the other is 1 1/2 sq. ft. What is the difference in their area size?

Solution.   Subtract 100 sq. in. from 1 1/2 sq. ft.

Step 1. 1 1/2 sq. ft. = 3/2 sq. ft.
Step 2. 3/2 x 144 = 216 sq. in.
Step 3. 216 sq. in. – 100 sq. in. = 116 sq. in.

Explanation. • First change 1 1/2 sq. ft. to an improper fraction (3/2) and multiply by 144 to obtain the total square inches in that portion (216 sq. in.). • Subtract 100 sq. in. from 216 sq. in., which equals 116 sq. in.

Multiplication

Multiplication of square units expressed in two or more denominations is similar to multiplication of linear units. Each unit of measure is multiplied separately. The smaller units are then changed to larger units by dividing by the proper number.

Another way to multiply square units of different denominations is to combine the smaller and larger units as a mixed number and multiply.

Another procedure is to express larger units in a smaller denomination and multiply. After this is done, the product is divided by the proper number.

Illustration. Multiply 5 sq. ft. 40 sq. in. by 8.

Solution. Step 1. 5 sq. ft. 40 sq. in. x 8 = 40 sq. ft. 320 sq. in.
Step 2. 320 ÷ 144 = 2 sq. ft. 32 sq. in.
Step 3. 40 sq. ft. + 2 sq. ft. 32 sq. in. = 42 sq. ft. 32 sq. in.

Explanation. • Multiply 40 sq. in. by 8 which equals 320 sq. in. and 5 sq. ft. by 8, equaling 40 sq. ft. Divide 320 sq. in. by 144 which equals 2 sq. ft. 32 sq. in. • Add 40 sq. ft. and 2 sq. ft. and annex 32 sq. in. • The answer is 42 sq. ft. 32 sq. in.

Illustration. Many carpenters install paneling, which is interior wall covering, as their fulltime occupation. A carpenter is planning to cover four walls with paneling. The walls measure 150 sq. ft. each. How many squares of paneling are required to cover all four walls?

Solution. (1 square = 100 sq. ft.)
Step 1. 50 sq. ft. = 50/100 = 1/2 square
Step 2. 1 1/2 x 4 = 3/2 x 4 = 12/2 = 6 squares

Explanation. • Follow the second procedure for multiplication of square units of different denominations. • Since 1 square is equal to 100 sq. ft., 50 sq. ft. is 1/2 square. • Annexing 1/2 square to 1 square equals 1 1/2 squares. Change 1 1/2 to an improper fraction (3/2) and multiply by 4. The answer is 6 squares.

Division

To divide square units when they are of different denominations, change the units to the smaller denomination and divide by the number. The quotient may then be changed back to its combined larger and smaller equivalents, such as feet and inches or yards and feet.

Illustration. A *pentagon* is a closed figure with five straight sides and five angles. The pentagon in figure 25-6 has a total of 16 sq. ft. 6 sq. in. of surface area divided equally among the five sides. How many square feet are there in each side? *Note: Surface area* may be defined as the area of the exposed part of an object.

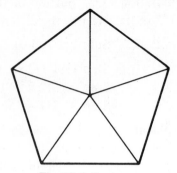

Fig. 25-6 Pentagon

Solution. Step 1. 144 sq. in. x 16 sq. ft. = 2304 sq. in.
Step 2. 2304 sq. in. + 6 sq. in. = 2310 sq. in.
Step 3. 2310 sq. in. ÷ 5 = 462 sq. in.
Step 4. 462 sq. in. ÷ 144 sq. in. = 3 sq. ft. 30 sq. in.

Explanation. • Add 16 sq. ft. (changed to 2304 sq. in.) to 6 sq. in. for a total of 2310 sq. in. • Divide 2310 sq. in. by the number of sides (5) for a quotient of 462. • The quotient is changed back to square feet by dividing by 144. • The answer is 3 sq. ft. 30 sq. in.

A short procedure is to divide each unit separately, beginning with the larger unit. If there is a remainder, change it to units of the next smaller denomiantion and add it. Then divide.

Illustration. Divide 16 sq. ft. 6 sq. in. by 5.

Solution. Step 1. 16 sq. ft. ÷ 5 = 3 sq. ft.; (1 sq. ft. remainder)
Step 2. 6 sq. in. + 144 sq. in. = 150 sq. in.
Step 3. 150 sq. in. ÷ 5 = 30 sq. in.

Explanation. • Divide 16 sq. ft. by 5, which leaves a quotient of 3 sq. ft. with 1 sq. ft. remaining. • Change the remaining square foot to inches (144) and add the 6 sq. in. from the original problem, giving 150 sq. in. • Divide 150 sq. in. by 5, which gives a quotient of 30 sq. in. • Annex this quotient (30 sq. in.) to 3 sq. ft. • The answer is 3 sq. ft. 30 sq. in.

APPLICATION

1. Convert each of the following square units of measurement as indicated.

   a. 6 sq. ft. to square inches
   b. 19 sq. ft. to square inches
   c. 24 sq. ft. to square inches
   d. 30 sq. ft. to square inches
   e. 3 sq. ft. to square inches
   f. 16 sq. ft. to square inches
   g. 50 sq. ft. to square inches
   h. 26 sq. ft. to square inches

2. Convert each of the following square units of measurement as indicated.

   a. 19 sq. in. to square feet
   b. 128 sq. in. to square feet
   c. 72 sq. in. to square feet
   d. 400 sq. in. to square feet
   e. 2,310 sq. in. to square feet
   f. 121 sq. in. to square feet
   g. 857 sq. in. to square feet
   h. 720 sq. in. to square feet
   i. 1,440 sq. in. to square feet
   j. 2,880 sq. in. to square feet

3. Change square yards to square feet as indicated.

   a. 15 sq. yd. to square feet
   b. 14 sq. yd. to square feet
   c. 23 sq. yd. to square feet
   d. 45 sq. yd. to square feet
   e. 90 sq. yd. to square feet
   f. 144 sq. yd. to square feet
   g. 847 sq. yd. to square feet
   h. 207 sq. yd. to square feet

4. Change square feet to square yards as indicated.

   a. 6 sq. ft. to square yards
   b. 12 sq. ft. to square yards
   c. 24 sq. ft. to square yards
   d. 48 sq. ft. to square yards
   e. 96 sq. ft. to square yards
   f. 18 sq. ft. to square yards
   g. 36 sq. ft. to square yards
   h. 144 sq. ft. to square yards
   i. 72 sq. ft. to square yards
   j. 288 sq. ft. to square yards

5. Convert the following square units as indicated.

   a. 2 sq. yd. to square inches   f. 1296 sq. in. to square yards
   b. 10 sq. yd. to square inches   g. 2592 sq. in. to square yards
   c. 15 sq. yd. to square inches   h. 5184 sq. in. to square yards
   d. 25 sq. yd. to square inches   i. 10368 sq. in. to square yards
   e. 6 sq. yd. to square inches   j. 5832 sq. in. to square yards

6. Convert the following square units as indicated.

   a. 200 sq. ft. to squares   f. 3 squares to square feet
   b. 250 sq. ft. to squares   g. 2 1/2 squares to square feet
   c. 400 sq. ft. to squares   h. 50 squares to square feet
   d. 460 sq. ft. to squares   i. 15 squares to square feet
   e. 1244 sq. ft. to squares   j. 9 1/4 squares to square feet

7. Add the following square units as indicated.

   a. 20 sq. in. plus 2 sq. ft.
   b. 150 sq. in. plus 5 sq. yd.
   c. 275 sq. ft. plus 7 1/2 sq. yd.
   d. 2 sq. ft. 5 sq. in. plus 3 sq. ft. 42 sq. in.
   e. 4 sq. yd. 1 sq. ft. plus 8 sq. ft. 30 sq. in.
   f. 4 squares plus 4 sq. yd.

8. Solve the following problems as indicated.

   a. Subtract 75 sq. ft. from 20 sq. yd.
   b. Multiply 2 sq. yd. by 6.
   c. Divide 8 sq. ft. 72 sq. in. by 4.
   d. Subtract 5 sq. yd. 1 sq. ft. from 8 sq. yd. 2 sq. ft.
   e. Multiply 10 1/2 squares by 5.
   f. Change 10 squares to square feet.

Unit 26
Area of Squares, Rectangles, Triangles, and Circles

OBJECTIVES

After studying this unit, the student should be able to

- find the area of a square, rectangle, triangle, circle, and semicircle.

- estimate material by using area measurement.

The area of a square, rectangle, triangle, or circle is the number of units of square measurement which make up its surface. To determine the area of a square or rectangle, multiply the width and length. The numbers must be of the same denomination when they are multiplied together, the area is expressed in square area.

FINDING THE AREA OF A SQUARE

To find the area of a square, the formula $A = S^2$ is used.

Illustration. Find the area of the square in figure 26-1.

Solution. Step 1. $A = (5)^2$
Step 2. $A = 5' \times 5' = 25$ sq. ft.

Explanation. • The formula $A = S^2$ is used because the length and width of a square have the same measurement. • The product of 5′ multiplied by 5′ is 25 sq. ft. • When a number is multiplied by itself, or squared, a small 2 is placed above and slightly to the right of the number.

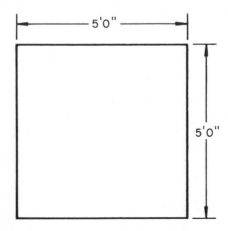

Fig. 26-1 Area of Square Sides

FINDING THE AREA OF A RECTANGLE

Usually the layout of a house or rooms in a house shows them as being rectangular in shape. The area of a rectangle is equal to the length multiplied by the width, and the area is in square measurement. The formula $A = lw$ is used when figuring the area of a rectangle.

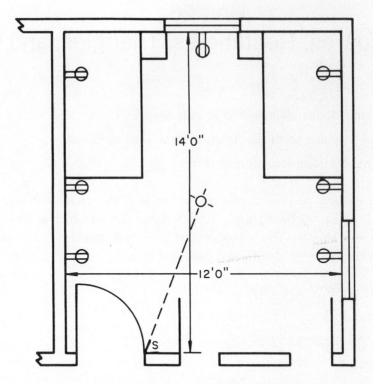

Fig. 26-2 Rectangular-shaped Room

Illustration. Find the area of the rectangular room layout in figure 26-2.

Solution. A = LW   A = 14′ x 12′ = 168 sq. ft.

Explanation. • Following the formula, the area is equal to the product of the length times the width. • The product of 14 multiplied by 12 is 168; the answer is expressed as 168 sq. ft.

FINDING THE AREA OF A TRIANGLE

There are various shapes of triangles. The area of a triangle is equal to one half the height multiplied by the base, or bottom: A = 1/2 hb. The result of the multiplication is expressed in square measurement.

Illustration. One example of triangular-shaped construction is shown in figure 26-3. Find the area of one end.

Solution.   Step 1. A = 1/2 x 12 x 8
                 Step 2. 6 x 8 = 48
                 Step 3. A = 48 sq. in.

Explanation. • According to the formula, multiply one half by the height or altitude (12″) and the base (8″). • The product is 48 sq. in., which is the amount of material needed to construct one end of the case, disregarding waste.

A *gable roof* is a ridged roof consisting of two sloping sections. A *gable* is the triangular-shaped end section which extends from the ridge to the eave, figure 26-4.

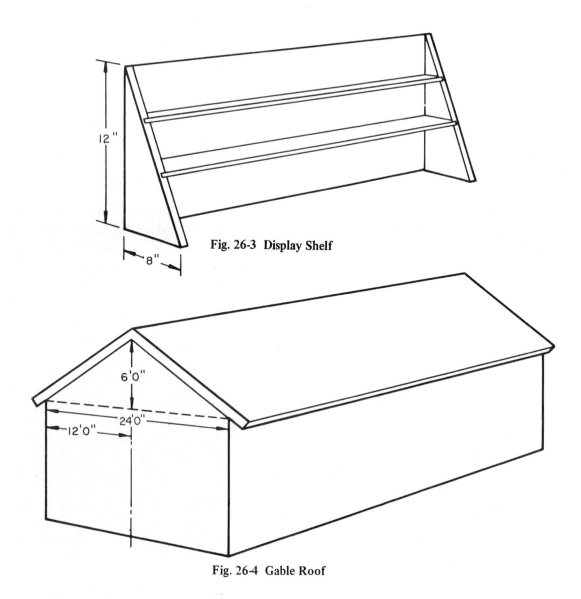

Fig. 26-3 Display Shelf

Fig. 26-4 Gable Roof

Illustration. The gable in figure 26-4 has a rise of 6' and a span of 24'. How many square feet of 1" x 8" shiplap are needed to cover both gables allowing 30% for waste?

Solution. a = rise 6'   b = span 24'
$1/2$ x 6 x 24 = 72
2 x 72 = 144
144 x .30 = 43.20
144 + 43 = 187 sq. ft.

Explanation. • Applying the formula, $1/2$ x 6 x 24 equals 72 sq. ft., which is the area of one gable. • Multiply 72 by 2, which equals 144 sq. ft., for the total area of both gables. • Multiply 144 by .30 for waste. The result is 43.20, which is rounded off to 43. • Add 144 and 43 to obtain the total number of square feet of shiplap needed to cover the gables, 187 sq. ft.

Formica is a laminated plastic product usually used for surface finishing. The amount of material required to cover a certain table top can be figured by the formula $A = \pi r^2$.

Illustration. The table top measures 4′ in diameter. How many square feet of formica are required to cover the top?

Solution.   (r = 1/2 diameter)

Step 1. r = 2
Step 2. r^2 = 2 x 2 = 4
Step 3. A = 3.1416 x 4 = 12.5664 sq. ft.

Explanation. • The area in square feet of this table top equals the product of 3.1416 times 4. These figures represent π and r^2. It takes 12.5664 sq. ft. to cover the top.

FINDING THE AREA OF A CIRCLE

A *circle* is a closed, curved plane with all points equidistant from its center. A circle always contains 360°.

To find the area of a circle, multiply π, or 3.1416, by r^2. (Remember that r is the radius; one half the diameter.)

Illustration. The power saw blade in figure 26-5 is 14″ in diameter. What is the area of the blade?

Solution.   r = 7 π = 3.1416

Step 1. r^2 = 7 x 7 = 49 sq. in.
Step 2. 3.1416 x 49 sq. in. =
153.9384 sq. in.

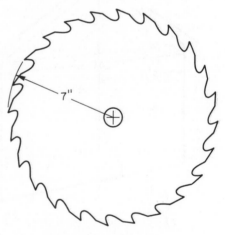

Explanation. • Substitute 3.1416 for π and 7^2 (49) for r^2 and multiply. • After multiplication, point off four decimal places to the left in the product. • The area of the power saw blade in figure 26-5 is 153.9384 or 154 sq. in. rounded off.

Fig. 26-5 Power Saw Blade

FINDING THE AREA OF A SEMICIRCLE

The area of a semicircle is equivalent to one half the area of a circle. If $A = \pi r^2$ is the formula for finding the area of a circle, then $A = 1/2\pi r^2$ would apply for a semicircle.

Illustration. A semicircle porch has a radius of 6′. What is the square area?

Solution.   Step 1. A = 1/2 x 3.1416 x (6 ft.)^2
Step 2. A = 1/2 x 3.1416 x 36 sq. ft.
Step 3. A = 56.5488 or 57 sq. ft.

Explanation. • Remember that the area of a semicircle is half the area of a circle. • Substitute 3.1416 for π, and 6^2 or 36 for r^2 (since the radius of the porch is 6′). • Multiply 1/2 x 3.1416 x 36. • The answer is 56.5488 or 57 sq. ft.

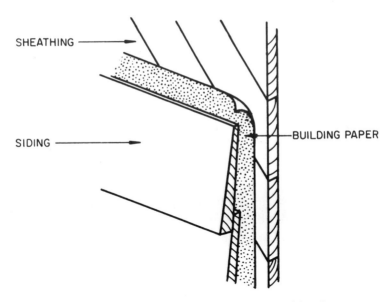

SHEATHING

SIDING

BUILDING PAPER

Fig. 26-6 Sheathing is Placed Under Siding and Building Paper.

ESTIMATING MATERIAL BY AREA MEASUREMENT

Many tradesmen estimate the amount of sheathing, insulating material, and other materials by first finding the area of the object.

Sheathing aids in the insulation of structures, figure 26-6. To determine the amount of sheathing needed to cover a certain wall, use the formula A = lw.

Illustration. How many pieces of 5/8" plywood which measure 4' x 8' are required to cover a wall 8' high and 52' long?

Solution.   Step 1. A = 52 x 8 = 416 sq. ft.
Step 2. 4 x 8 = 32 sq. ft. of sheathing in one sheet.

$$\frac{416 \ \text{(area to be covered)}}{32 \ \text{(number of sq. ft. in one sheet)}} = 13$$

Explanation. • Find the total area to be covered by multiplying length by width, which gives 416 sq. ft. • Find the number of square feet that one piece of plywood will cover by multiplying the length of the board by the width. • Divide the total area (416) by the square area of one piece (32). The answer is 13 sheets. • If the studs are correctly spaced and if there are no windows, no sawing is necessary. If sawing is required, allow one percent for waste.

ESTIMATING MATERIAL BY THE SQUARE

Roofing materials are figured by the square (100 sq. ft.). To determine the amount of roofing material needed to cover a roof, first determine the total surface area to be covered. This is done by multiplying the length by the width: A = lw/100. The product is in square measurement; usually in square feet. After the total area of roof surface has been determined, divide it by 100 sq. ft.

Illustration. If the total length of a rafter measures 11' 6" and the length of the roof is 48', how many squares of roofing material are required to cover a gable roof? Allow 12% for waste.

Solution. Step 1. A = 11 1/2' x 48'
Step 2. A = 23/2 x 48 = 552 sq. ft.
Step 3. .12 x 552 = 66.24 (waste)
Step 4. 66 + 552 = 618 sq. ft.
Step 5. Squares = 618/100 = 6.18
6.18 x 2 = 12.36 = 12 1/2 squares

Explanation. • Determine the total number of square feet for one side of the roof by multiplying 11 1/2' by 48', and multiplying that product by .12. • Add the two products. • The total number of square feet is 618. • Divide 618 by 100, which gives 6.18 squares to cover one side of the roof. • Multiply 6.18 by 2 which equals 12.36 squares. • Round off the answer to 12 1/2 squares to cover both sides of the roof.

APPLICATION

1. Find the area of squares with sides of the following measurements.

   a. 10' c. 52" e. 3' 4" g. 5 yd.
   b. 15' d. 33 1/2" f. 5' 6" h. 5 yd. 2'

2. Find the area of circles with the following diameter measurements.

   a. 4" e. 3.125' i. 33' l. 3/4'
   b. 5.4' f. 19' j. 43' m. 130 yd.
   c. 14' g. 24 yd. k. 9 1/2' n. 5' 6"
   d. 75' h. 5.8"

3. Find the areas of the figures as indicated.

   a. rectangle, 21' by 12'
   b. circle, 5' in diameter
   c. triangle, altitude 14"; base 9"
   d. 12' square
   e. 4 sides, 13' each
   f. rectangle, 2' by 3'
   g. circle, 7 1/2' in diameter
   h. triangle, altitude 16"; base 12"
   i. 4 sides, 4 1/2" each
   j. rectangle, 8.8" by 9"
   k. circle, 11' in diameter

4. What is the area of a square whose sides measure 12' 8"?

5. What is the area of a rectangle if its length measures 13' and width 11'?

6. Find the cost of each of the following.

   a. a piece of linoleum 10" by 16" at $.95 per sq. ft.
   b. a concrete sidewalk finish surface 4' wide 160' long at $3.55 per square yard.
   c. Roofing shingles installed on a roof 12' by 42' at $5.75 per square.

7. A room is 12' wide and 16' long. It has one door 3' by 7' and one picture window 8' by 5'. How many pieces of plasterboard which measure 4' x 8' are required to cover the walls allowing 10% for waste?

8. A certain roll of felt is 3' wide and 100' long. How many squares are there in the roll?

9. A room 25' by 30' is covered with linoleum which costs $0.75 per square foot. What is the total cost?

10. According to the dimensions of the drawing in figure 26-7, a is 10' and b is 10'. How many squares of roofing are required to cover both sides of the triangular-shaped section of the roof if 15% is allowed for waste?

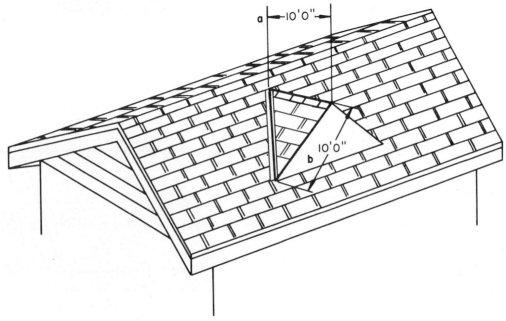

Fig. 26-7

11. Subflooring of structures is nailed directly against the floor joist. If a floor measures 27' 8" wide and 68 1/2' long, how many square feet of subflooring material are required? (Allow 29% for waste.)

12. An outdoor bandstand is constructed of bricks with a concrete floor. The floor is 21' in diameter. How many square feet of floor area are in the bandstand?

Unit 27
Area of Parallelograms and Trapezoids

OBJECTIVES

After studying this unit, the student should be able to

- find the area of parallelograms and trapezoids.

- estimate material for objects in the shape of parallelograms and trapezoids.

- determine the weight of objects in the shape of parallelograms and trapezoids.

Parallelograms and trapezoids are frequently used in carpentry layout. At times, the carpenter must determine the area for these objects.

FINDING THE AREA OF A PARALLELOGRAM

A *parallelogram* has four sides with opposite sides parallel and equal, figure 27-1. Notice that a rectangle can be classified as a parallelogram. The area of a parallelogram is equal to the height (h) multiplied by the base (b): $A = hb$. *Note:* The height is measured at right angles (90°) to the base.

Illustration. Find the area of the parallelogram in figure 27-1.

Solution.   h = 15″   b = 20″
           A = 15″ x 20″
           15″ x 20″ = 300 sq. in.

Explanation. • The area of the parallelogram in square inches is equal to the product of the height (15″) and base (20″). • The product of 15 and 20 is 300. • Therefore, the area of the parallelogram is 300 sq. in. or 2 1/12 sq. ft.

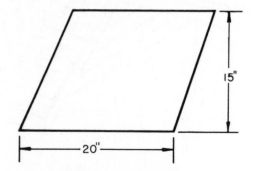

Fig. 27-1 Parallelogram

Hip and valley rafters extend diagonally from plate to ridge. When they are erected on the same plate and ridge, a parallelogram is formed, figure 27-2.

Illustration. How many squares of roofing are required to cover the area in figure 27-2? Allow 20% for waste because of end cuts.

Solution.   h = 12′ 4″   b = 20′
           Step 1.   A = 12′ 4″ x 20′
           Step 2.   A = 12 1/3 x 20

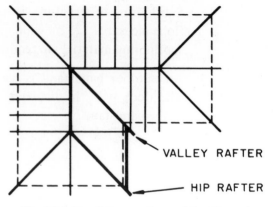

Fig. 27-2 Parallelogram Formed By Hip and Valley Rafters

Step 3. A = 37/3 x 20 = 740/3 = 246.66 sq. ft.
Step 4. 246.66 x .20 = 49.33
Step 5. 246.66 + 49.33 = 295.99 or 296 sq. ft.
Step 6. 296 ÷ 100 = 2.96 or 3 squares

Explanation. • First, find the area of the parallelogram formed in the figure by multiplying 12′ 4″ by 20′. • Change 12′ 4″ to 12 1/3′, or 37/3. • The product of 37/3 times 20 is 246.66 sq. ft. • To determine the percentage of waste, change 20% to .20 and multiply by 246.66. • Add the product (49.33) to 246.66. • Round off the sum (295.99) to the nearest hundredth (296). • To find the number of squares of material needed for the area, divide 296 sq. ft. by 100. • The quotient is 2.96, or 3, squares.

FINDING THE AREA OF A TRAPEZOID

A *trapezoid* has four sides, two sides which are parallel and two sides which are not parallel, figure 27-3. The area of a trapezoid is equal to the height multiplied by the average of the two parallel sides: $A = h \times [(t+b)/2]$. The formula can also be expressed as $1/2 (t+b) \times h$. The t represents the length of the top side; b, the length of bottom or base; and h, the height.

Illustration. The top side of the trapezoid in figure 27-3 is 5′ long. The bottom side is 9′ long and the height is 6′. What is the area in square feet?

Solution. $A = h \times \dfrac{t+b}{2}$

Step 1. $A = 6 \times \dfrac{5+9}{2}$

Step 2. $A = 6 \times \dfrac{14}{2} = \overset{3}{\cancel{6}} \times \dfrac{14}{\cancel{2}_{1}}$

$A = 42$ sq. ft.

   or

$A = 1/2 (t + b) \times h$

Step 1. $A = 1/2 (5 + 9) \times 6$

Step 2. $A = 1/2 \times 14 \times 6 = 1/\cancel{2} \times \overset{7}{\cancel{14}} \times 6 = 42$

$A = 42$ sq. ft.

Explanation. • Following the first formula, multiply the height, (6′) by the average of the top side (5′) and bottom side (9′). • To find the average, divide the sum of 5 and 9 (14) by 2. • Multiply the product (7) by the height (6). • The area of the trapezoid is 42 sq. ft. • The procedure in the second formula is very similar to the first, except that 1/2 is used to determine the average of the two sides.

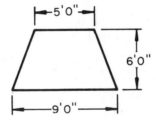

Fig. 27-3 Trapezoid

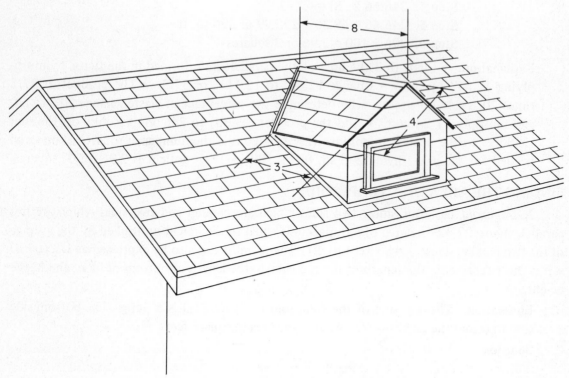

Fig. 27-4 Dormer with Trapezoidal Section

The dormer in figure 27-4 is in the shape of a trapezoid. A *dormer* is that section of a roof which projects above the main roof surface to allow space for a window unit. Notice that part of the dormer in the figure forms a trapezoid.

Illustration. A carpenter must determine the area of the trapezoid in the figure to purchase necessary materials. With given dimensions, how many square feet of roofing material are required to cover the dormer? Allow 15% for waste due to end cuts.

Solution. $A = 4 \times \dfrac{8 + 3}{2}$

$A = \overset{2}{\cancel{4}} \times \dfrac{11}{\underset{1}{\cancel{2}}} = 22$ sq. ft.

22 x 2 = 44 sq. ft.

44 x .15 = 6.60

44 + 6.60 = 50.60 = 51 sq. ft.

Explanation. • The amount of roofing required to cover the dormer is determined by multiplying the height (4′) by the average of the top side (8′) and bottom side (3′). • Then multiply the product (22) by 2 to find the amount it takes for both slopes on the dormer (44). • Multiply the total area (44 sq. ft.) by the percentage of waste (15%). • Add this amount (6.60) to the total area (44 sq. ft.) to find the total square feet of material needed to cover the dormer (51 sq. ft.).

APPLICATION

1. Find the area of parallelograms with the following measurements.

   a. h = 4′, b = 8′   f. h = 4″, b = 8″
   b. h = 4 yd., b = 8 yd.   g. h = 4.4″, b = 8.8″
   c. h = 6 1/2′, b = 12 1/2′   h. h = 8.5″, b = 16.5′
   d. h = 8.5′, b = 16.5′   i. h = 9′ 10″, b = 20′ 4″
   e. h = 6′ 8″, b = 12′ 9″

2. Find the area of trapezoids with the following measurements.

   a. t = 10″, b = 15″, h = 8″   d. t = 18.8″, b = 36.2″, h = 24.10″
   b. t = 12′, b = 24′, h = 14′   e. t = 80′, b = 160′, h = 40′
   c. t = 8.8′, b = 14.10′, h = 7.5′

3. What is the area of a parallelogram with a height of 8′ and a base of 12′?

4. Find the area of a trapezoid with the following measurements: top side, 6′ 6″; base, 12′; and height, 9 1/2′.

5. The hip roof in figure 27-5 is shaped like a trapezoid. Sheathing, nailed on the roof, must then be covered with roofing material. Determine the amount of sheathing material and the amount of asphalt shingles (in square measurement) needed to cover the trapezoidal sections. Allow 33% waste for the sheathing and 10% for the shingles.

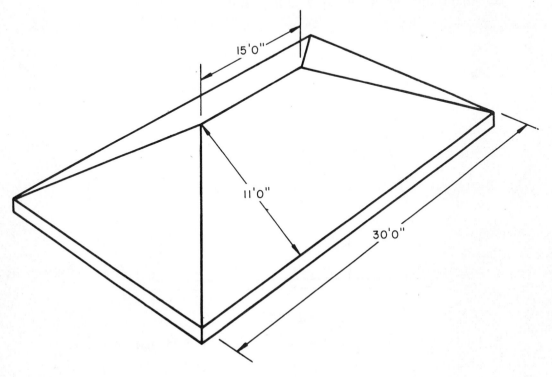

Fig. 27-5

6. The base of the parallelogram in figure 27-6 is 24″ and the height is 18″. How many square feet of metal are required to cover the two sides?

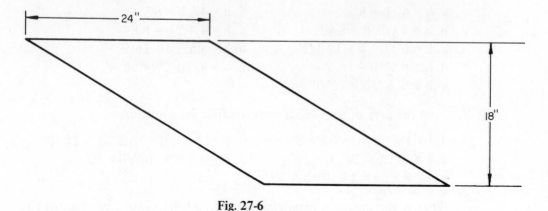

Fig. 27-6

7. A carpenter must measure a concrete surface in square yards before finishing it. The plot is shaped like a trapezoid with the sides measuring 30′ and 20′ and the height measuring 50′. How many square yards of surface area are there to finish?

8. What is the area of a trapezoid if the height is 19″ and the parallel sides are 11″ and 15″ long?

9. What is the area of the trapezoid in figure 27-7?

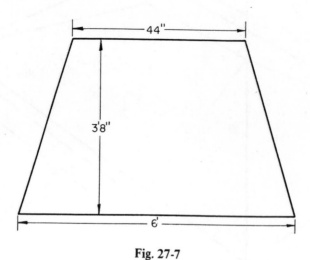

Fig. 27-7

10. Find the area of a parallelogram with a height of 25′ and a base of 45′.

11. Find the area of a parallelogram with a height of 15 yd. 2′ and a base of 25 yd. 2′.

12. Find the area of a trapezoid with a top side of 4.8', a base of 4.2', and a height of 2 3/4'.

13. Find the area of a parallelogram with a height of 10" and a base of 32".

14. Find the area of a trapezoid with a top side measuring 43", a base measuring 53", and height measuring 10".

15. What is the area of a trapezoid if the height is 8" and the parallel sides measure 9" and 16"?

Unit 28
Area of Rectangular Solids

OBJECTIVES

After studying this unit, the student should be able to

- determine the area of rectangular solids and cubes.
- determine the weight and cost of objects by total surface area.

Rectangular solids and cubes are three dimensional objects having width, height, and depth. In a *rectangular solid,* four of the object's surfaces are rectangles. In a *cube,* all six surfaces are equal squares. An example of a rectangular solid is a brick; any perfectly square, six-sided object is a cube.

Many jobs in carpentry involve formulas which determine surface area of rectangular solids and cubes, such as estimating materials.

FINDING THE AREA OF A RECTANGULAR SOLID

The rectangular solid, figure 28-1, has six surfaces; all surfaces must be counted when determining the total area. The recommended formula to use is A = 2 lw + 2 lh + 2 wh.

Illustration. Find the total area of the rectangular solid in figure 28-1 whose length is 8″, width 5″, and height 6″.

Solution.　　Step 1.　A = 2 x 8 x 5 +
　　　　　　　　　　　　　　2 x 8 x 6 +
　　　　　　　　　　　　　　2 x 5 x 6
　　　　　　　　Step 2.　A = 80 + 96 + 60
　　　　　　　　Step 3.　A = 236 sq. in.

Explanation. The total surface area of a rectangular solid is equal to the sum of 2 times the product of the length and width, 2 times the product of the length and height, and 2 times the product of the width and height; in this problem, 2 x 8 x 5 = 80, 2 x 8 x 6 = 96, and 2 x 5 x 6 = 60. Add the products 80, 96, and 60. The total surface area of the rectangular solid is 236 sq. in.

Illustration. Find the amount of wood required to construct outside surfaces of the cabinet in figure 28-2 whose length is 2′ 4″, width 1′ 3″, and height 2′ 5″.

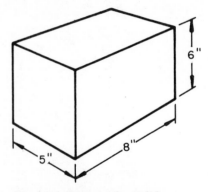

Fig. 28-1　Rectangular Solid

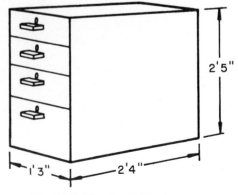

Fig. 28-2　Wooden Cabinet

Solution. A = 2 lw + 2 lh + 2 wh

    Step 1. A = (2 x 2 1/3 x 1 1/4) + (2 x 2 1/3 x 2 5/12) + (2 x 1 1/4 x 2 5/12)

    Step 2. A = (2 x 7/3 x 5/4) + (2 x 7/3 x 29/12) + (2 x 5/4 x 29/12)

    Step 3. $\dfrac{\overset{1}{\cancel{2}}}{1} \times \dfrac{7}{3} \times \dfrac{5}{\underset{2}{\cancel{4}}} = \dfrac{35}{6} = 5\dfrac{5}{6}$

    Step 4. $\dfrac{\overset{1}{\cancel{2}}}{1} \times \dfrac{7}{3} \times \dfrac{29}{\underset{6}{\cancel{12}}} = \dfrac{203}{18} = 11\dfrac{5}{18}$

    Step 5. $\dfrac{\overset{1}{\cancel{2}}}{1} \times \dfrac{5}{4} \times \dfrac{29}{\underset{6}{\cancel{12}}} = \dfrac{145}{24} = 6\dfrac{1}{24}$

    Step 6. A = 5 5/6 + 11 5/18 + 6 1/24

       = 5 60/72 + 11 20/72 + 6 3/72 = 22 83/72 = 23 11/72 sq. ft.

Explanation. Change the measurements to mixed numbers and then to improper fractions. Multiply according to the formula and add the three products. The result (23 11/72) may be rounded off to a whole number for the final answer, 24 sq. ft.

The total surface area of a room (all six sides) can be found by using the formula for finding the area of a rectangular solid.

Illustration. A certain room is 11′ by 16′, with a wall height of 8′. Find the total surface area.

Solution.   A = (2 x 16 x 11) + (2 x 16 x 8) + (2 x 11 x 8)

      A = 352 + 256 + 176

      A = 784 sq. ft.

Explanation. According to the formula, the two sides which measure 16′ by 11′ equal 352 sq. ft. The two sides which measure 16′ by 8′ equal 256 sq. ft., and the other 2 sides, 11′ by 8′, equal 176 sq. ft. The sum of 352 sq. ft., 256 sq. ft., and 176 sq. ft. (784 sq. ft.) is the total surface area of the room.

FINDING THE AREA OF A CUBE SOLID

A *cube solid* is an object which has six equal sides with all corners having right angles. The cube solid is similar to the rectangular solid except that all sides are equal on the cube solid.

The formula to use for determining the total outside surface area of a cube solid is A = 6S². The 6 represents the number of surfaces and the S² indicates that the length of each side is multiplied by itself. Figure 28-3 shows the bounding lines of a cube solid.

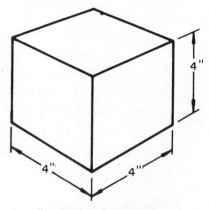

Fig. 28-3 Cube Solid

Illustration. A *brick pier* is masonry which acts as a support, figure 28-4. Find the surface area of the brick pier with measurements as shown.

Solution. $A = 6S^2$

Step 1. $A = 6 \times 16^2$

Step 2. $A = 6 \times 256 =$
$\qquad\qquad\qquad 1536$ sq. in.

Explanation. • First multiply 16×16 which equals 256 sq. in. and then multiply by the number of sides (6). • The product is 1536 sq. in., which represents the total surface area of the cube solid.

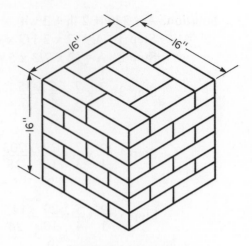

Fig. 28-4 Brick Pier

Illustration. Find the surface area in square inches of a steel gas tank whose cube sides measure 18 1/2″ each.

Solution. $A = 6S^2$

Step 1. $A = 6 \times (18\ 1/2)^2$

Step 2. $6 \times 18\ 1/2 \times 18\ 1/2\ =$

Step 3. $\overset{3}{\cancel{6}} \times \dfrac{37}{\underset{1}{\cancel{2}}} \times \dfrac{37}{2} = \dfrac{4107}{2}$

$A = 2053\ 1/2$ sq. in.

Explanation. • Apply the formula, $A = 6 \times 18\ 1/2^2$. • Change $18\ 1/2 \times 18\ 1/2$ to $37/2 \times 37/2$ and multiply by 6. • The total surface area of the gas tank is 2053 1/2 sq. in.

APPLICATION

1. Find the total area of the following rectangular solids.

   a. Length 10′, width 14′, and height 8′
   b. Length 14′, width 12′, and height 9′
   c. Length 5.5′, width 6′ 6″, and height 6′
   d. Length 2 yd. 2′, width 2 yd. 1′, and height 3 yd.

2. Find the area of each of the following as indicated.

   a. 20″ by 20″ by 10″
   b. 15′ 6″ by 12′ 4″ by 8′
   c. 14′ 3″ by 10′ 6″ by 9′
   d. 11′ 9″ by 11′ 9″ by 10′
   e. 29′ by 72′ by 11′

3. Find the total surface area of each of the cube solids.

a. 12″	d. 24″	g. 10′	j. 21′	m. 9.1′
b. 9″	e. 32″	h. 8′	k. 5.5′	n. 10.8′
c. 15″	f. 5′	i. 18′	l. 7.6′	o. 12.4′

4. Find the total surface area of each of the rectangular solids with the following dimensions.

	a	b	c	d	e
Length	14'	16'	32"	6.6'	10 1/2'
Width	16'	15'	24"	5.2'	12 1/2'
Height	8'	8'	20"	4.8'	6 1/2'

5. How many square feet of plywood are needed to construct the box in figure 28-5?

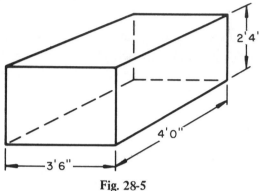

Fig. 28-5

6. A certain room in a home is cube-shaped, with all dimensions measuring 8'. What is the total surface area (including walls, ceiling, and floor) of the room? Do not consider windows and doors in your answer.

7. A study in a home is a rectangular solid 2' by 12', with the walls measuring 8' high. The room has one window 3' by 4' and one door 2 2/3' by 6 2/3'. Find the total area of the room deducting the area for the window and door.

8. How many square feet of 5/8" plywood are required to construct a rectangular solid 4' by 4' by 8'?

9. What is the cost to paint a rectangular solid 10' by 20' by 8', at 65¢ per square foot?

10. A speaker stand is 24" wide, 24" deep and 40" high and has a top. How many square feet make up the speaker stand? How much will it cost to sand and varnish the stand at 25¢ per square foot?

11. Determine the amount and weight of material required to construct a cabinet 6' long, 2' wide and 6' high. The material weighs 3/4 lb. per square foot. (Do not allow for waste or joints.)

12. Determine the weight of a 15" cube solid made of metal which weighs 1 1/4 lb. per square foot.

13. A certain bedroom is a 14' cube solid. What is the surface area of the room?

14. Find the surface area of a cube solid whose sides measure 16'.

15. Find the cost of constructing a cabinet whose cube sides measure 20" at 3¢ per square inch.

Unit 29
Lateral Area and Total Area of Cylinders

OBJECTIVES

After studying this unit, the student should be able to

- find the lateral area and total area of a cylinder.
- determine the weight and building cost of a cylinder before it is constructed.

A *cylinder* is a circle of a specified diameter in space with a straight vertical line moving continuously around the circle to form a closed plane figure, figure 29-1. Many cylindrical objects are commonly used in carpentry, such as rods, dowels, and pipes.

A cylinder has length, sometimes called height or *altitude*. Each end of the cylinder is a *base*, and the diameter of the ends is considered the diameter of the cylinder.

The lateral area of a cylinder includes only the outside curved surface. The total area includes the outside curved surface and the area of the two bases.

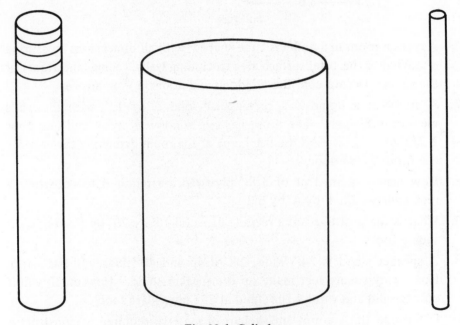

Fig. 29-1 Cylinders

FINDING THE LATERAL AREA OF A CYLINDER

To find the lateral area of a cylinder, use the formula $A = \pi dh$. When the radius is given instead of the diameter, the formula $A = 2\pi rh$ is used.

Illustration. Find the lateral area of the cylinder in figure 29-2.

Solution. $A = \pi dh$
             Step 1. $A = 3.1416\ (\pi) \times 4'' \times 6''$
             Step 2. $A = 75.3984$ sq. in.

Explanation. • The circumference is found by multiplying 3.1416 by 4 which equals 12.5664. • The lateral area is then found by multiplying 12.5664 by 6 which equals 75.3984.

Material needed for construction of an object can be determined by finding the lateral area of the object.

Illustration. Determine the amount of material needed for constructing an object with a radius of 6″ and a height of 4′. Do not include the bases in the problem.

Fig. 29-2 Cylinder

Solution.   A = 2π rh   6″ = .5′
  A = 2 x 3.1416 x .5 x 4 = 12.5664 sq. ft.

Explanation. • Since the radius is given instead of the diameter, the formula A = 2πrh is used. • Multiply 2 times 3.1416 and multiply that product (6.2832) by .5′ (changed from 6″). • The product at this point (3.1416) gives the distance around the object which is used to determine the length of the piece. • The lateral area is found by multiplying 3.1416 by 4. • The answer is 12.5664 sq. ft.

FINDING THE TOTAL AREA OF A CYLINDER

The total area of a cylinder is found by applying one of two formulas: A = 2π rh + 2π r^2 or A = π dh + π d^2/2.

The lateral area is found by applying the formula A = π dh; π d^2/2 is used to find the area of the two bases. The sum of the two areas is the total area of the cylinder.

Illustration. Figure 29-3 shows a cylinder which has a diameter of 6″ and a height of 8″. What is the total area of the cylinder?

Solution.   A = π dh + πd^2/2
  Step 1. A = 3.1416 x 6 x 8 +
      3.1416 x 36/2
  Step 2. 150.7968 + 56.5488 =
      207.3456 sq. in.

Explanation. • The product of 3.1416 x 6 x 8 gives the lateral area. • The product of 3.1416 x 36 divided by 2 gives the area of the two bases. • The sum of the two (150.7968 and 56.5488) is the total area of the cylinder, 207.3456 sq. in.

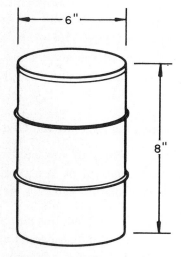

Fig. 29-3 Cylinder with Bases

Illustration. Find the total area of a cylinder with a radius of 5′ and a height of 25′.

Solution.   A = 2πrh + 2πr²
Step 1. A = 2 x 3.1416 x 5 x 25 + 2 x 3.1416 x 5²
Step 2. 785.40 + 157.080 = 942.48 sq. ft.

Explanation. • Determine the lateral area of the cylinder (785.40 sq. ft.) by multiplying 2 times pi times the radius times the height. • Determine the area of the two bases (157.080 sq. ft.) by multiplying 2 times pi times the radius squared. • The sum of the lateral area and the area of the two bases (942.48) is the total area of the cylinder.

DETERMINING WEIGHT AND COST

The weight and building cost of a cylinder can be determined after the square area is known.

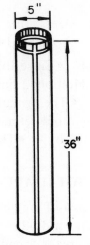

Illustration. The pipe in figure 29-4 is 36″ high and 5″ in diameter. It is made of metal weighing 3/4 lb. per square foot which costs 80¢ per pound. What is the cost of the metal in the construction of the pipe, allowing 3/4 sq. ft. for waste?

Solution.   A = π dh
Step 1. A = 3.1416 x 5 x 36 =
565.488 sq. in.
Step 2. 565.488 ÷ 144 = 3.927
sq. ft.
Step 3. 3.927 + .75 = 4.677
sq. ft. (total area)
4.677 x .75 = 3.50775 lb.
Cost = 3.50775 x .80 = $2.81

Fig. 29-4 Cylindrical Pipe

Explanation. Determine the lateral area of the pipe by multiplying 3.1416 by 5 by 36, which equals 565.488 sq. in. • Divide 565.488 sq. in. by 144, which equals 3.927 sq. ft. • Change 3/4 sq. ft. to .75 and add it to 3.927, which equals 4.677 sq. ft., the total area of the material. • Multiply 4.677 by .75 lb. for the weight of the material; 3.50775 lb. • Finally, multiply 3.50775 x .80. This gives the cost of the material for the pipe, $2.81.

Determining the Amount and Cost of Insulation

The efficiency of air-conditioning and heating systems depends to a large extent upon an adequate amount of insulation in such places as ceilings, walls, and around pipes. The amount of insulation and cost for labor can be determined by finding the area of the pipe.

Illustration. Find the number of square feet of material and the labor cost for the insulation of a heating and air-conditioning pipe measuring 6″ in diameter and 120′ in length. Labor cost is 20¢ per square foot.

Solution. $A = \pi\,d\,l$

      Step 1. A = 3.1416 x .5 x 120 = 188.49600

      Step 2. A = 188.50 sq. ft.

      Cost = 188.50 x .20 = $37.70

Explanation. • Change 6″ to .5′. • Multiply .5′ by pi (3.1416) by 120′, which equals 188.50. • Multiply total square feet (188.50) by the cost per foot (20¢) for the total cost for insulation, $37.70.

APPLICATION

1. Find the total area of cylinders with the following dimensions.

    a. radius 2″, height 6″   f. radius 2′, height 4′

    b. diameter 2″, height 6″   g. diameter 2′, height 4′

    c. radius 2′, height 6′   h. radius 2′ 6″, height 9′ 6″

    d. diameter 1 yd., height 2 yd. 2′   i. diameter 1 1/2′, height 12 1/2′

    e. radius 1/2 yd., height 2/3 yd.   j. radius .5 yd., height .8 yd.

2. Find the lateral area of cylinders with the following dimensions.

    a. radius 2″, height 6″   f. radius 3′, height 9′

    b. radius 4″, height 12″   g. radius 30″, height 80″

    c. radius 2 1/2′, height 8′ 6″   h. radius 2 1/3′, height 10′ 4″

    d. diameter 2.2″, height 4.6″   i. diameter 3.4′, height 3.4′

    e. diameter 1 yd., height 15 yd.

3. The soil pipe in figure 29-5 is 6″ in diameter and 32″ long. Find the lateral area of the plastic pipe.

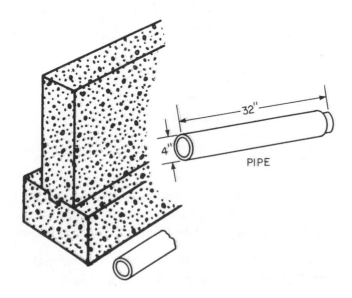

Fig. 29-5

4. How many square yards are there in 5 round columns which are each 18″ in diameter and 20′ high?

5. How many square feet of sheet metal are there in a tank which has a diameter of 3′ 4″ and a height of 4′ 4″? Allow 2 inches for a lap joint.

6. The outside diameter of the conduit pipe in figure 29-6 is 5/8″ and the length is 64″. If the pipe weighs .05 lb. per square inch, what is the total weight of the pipe?

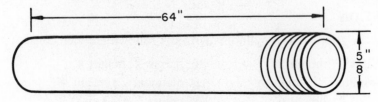

Fig. 29-6

7. The *flue* of a chimney carries off smoke, air, and gases from fire. What is the total measurement (in square inches) of the flue in figure 29-7?

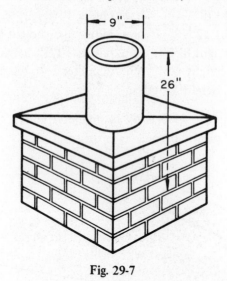

Fig. 29-7

8. What is the lateral area of a cylinder which has a diameter of 1.125″ and a length of 22.750 inches?

9. Find the number of square feet of insulating material and the labor cost for the insulation of heating and air-conditioning pipe measuring 8″ in diameter and 90′ in length. The labor cost is 25¢ per square foot.

10. Find the cost of insulating the total area of 5 water tanks measuring 18″ in diameter and 4′ in length at 25¢ per square foot.

11. Find the cost of insulating the total area of a water tank 9′ in diameter and 9′ in height at 55¢ per square yard.

Unit 30   Using the English System: Cubic Measurement

OBJECTIVES

After studying this unit, the student should be able to

- change cubic units of one denomination to another denomination.
- add, subtract, multiply, and divide cubic units of different denominations.

Cubic measurement differs from linear and square measurement in that it includes length, width, and thickness. These three dimensions are often used by the carpenter to determine the volume of solids.

Volume measurement equivalents commonly used in carpentry include:

$$1 \text{ cubic foot (cu. ft.)} = 1728 \text{ cubic inches (cu. in.)}$$
$$1 \text{ cubic yard (cu. yd.)} = 27 \text{ cubic feet (cu. ft.)}$$

Note: Cubic measurement refers to volume or capacity.

CHANGING CUBIC FEET TO CUBIC INCHES

To change cubic feet to cubic inches, multiply the number of cubic feet by 1,728.

Illustration. If a container holds 2 cu. ft. of water, how many cubic inches can the container hold?

Solution.   2 cu. ft. x 1,728 cu. in. = 3,456 cu. in.

Explanation. • The process involves only multiplication. • Since 1,728 cu. in. equals one cu. ft., multiply 2 times 1,728 which equals 3,456 cu. in.

CHANGING CUBIC INCHES TO CUBIC FEET

To change cubic inches to cubic feet, divide the number of cubic inches by 1,728.

Illustration. If a container holds 3,456 cubic inches of water, how many cubic feet does it contain?

Solution.   3,456 cu. in. ÷ 1,728 cu. in. = 2 cu. ft.

Explanation. • To change cubic inches to cubic feet, divide the number of cubic inches by 1,728, which equals 2 cu. ft.

CHANGING CUBIC YARDS TO CUBIC FEET

Many times, carpenters must work with objects which are measured in cubic yards, such as concrete foundations and sidewalks. To carry out addition, sometimes it is necessary to convert the measurement to a lower denomination.

To change cubic yards to cubic feet, multiply the number of cubic yards by 27.

Illustration. If a container holds 2 cubic yards of water, how many cubic feet does it contain?

Solution.   2 cu. yd. x 27 cu. ft. = 54 cu. ft.

Explanation. • An area which measures 3' x 3' x 3' is equal to 27 cu. ft. and 27 cu. ft. equals one cubic yard. • A container that is twice the size holds 54 cu. ft. • Since 1 cu. yd. equals 27 cu. ft., 2 cu. yd. equal 54 cu. ft. • To solve the problem, multiply 2 by 27 cu. ft. • The product is 54 cu. ft.

CHANGING CUBIC FEET TO CUBIC YARDS

To change cubic feet to cubic yards, divide the number of cubic feet by 27.

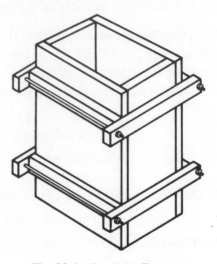

Illustration. • *Concrete forms* are metal or wooden molds into which concrete is poured, figure 30-1.

Concrete required to fill a certain concrete form is 81 cu. ft. Change 81 cu. ft. to cubic yards.

Solution.   81 cu. ft. ÷ 27 cu. ft. = 3 cu. yd.

Explanation. • When changing cubic feet to cubic yards, 27 is the conversion factor. • Divide the number of cubic yards, in this case 81, by the conversion factor which is 27. • The quotient is 3 cu. yd.

Fig. 30-1 Concrete Form

CHANGING CUBIC YARDS TO CUBIC INCHES

To change cubic yards to cubic inches, first change cubic yards to cubic feet by multiplying the number of cubic yards by 27. Then change the cubic feet to cubic inches by multiplying the number of cubic feet by 1,728.

Illustration. Change 1 1/2 cu. yd. to cubic inches.

Solution.   Step 1. 1 1/2 cu. yd. x 27 cu. ft. = 3/2 cu. yd. x 27 cu. ft. = 40 1/2 cu. ft.
                  Step 2. 40 1/2 cu. ft. x 1,728 cu. in. = 81/2 cu. ft. x 1,728 = 69,984 cu. in.

Explanation. • Multiply 1 1/2 cu. yd. by the number of cubic feet in 1 cu. yd. (27 cu. ft.). • The product is 40 1/2 cu. ft. Change cubic feet to cubic inches by multiplying 40 1/2 cu. ft. by 1,728. • The product (69,984 cu. in.) is the number of cubic inches in 1 1/2 cu. yd.

CHANGING CUBIC INCHES TO CUBIC YARDS

To change cubic inches to cubic yards, first change cubic inches to cubic feet by dividing the number of cubic inches by 1,728. Then change cubic feet to cubic yards by dividing the number of cubic feet by 27.

Illustration. The container in figure 30-2 holds 46,656 cu. in. How many cubic yards does it hold?

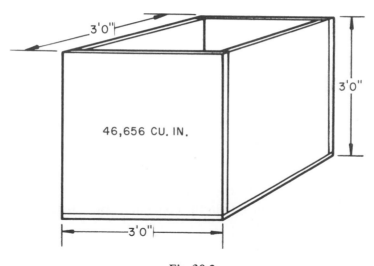

Fig. 30-2

Solution. Step 1. 46,656 cu. in. = 46,656 cu. in./1,728 cu. in. = 27 cu. ft.
Step 2. 27 cu. ft. ÷ 27 cu. ft. = 1 cu. yd.

Explanation. • Change 46,656 cu. in. to cubic feet by dividing 46,656 by 1,728, equaling 27 cu. ft. Then divide 27 cu. ft. by 27. The quotient is 1 cu. yd.

FUNDAMENTAL PROCESSES IN CUBIC MEASUREMENT

Addition

Illustration. The concrete form in figure 30-3 is in two parts. One part holds 1 cu. ft. 864 cu. in. of concrete; the other part holds 2 cu. ft. 864 cu. in. What is the total amount of concrete the form holds in cubic feet?

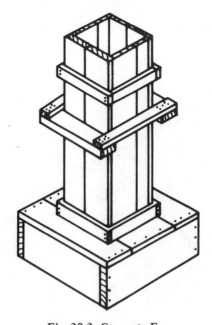

Fig. 30-3 Concrete Form

Solution.   Step 1. 1 cu. ft. 864 cu. in. + 2 cu. ft. 864 cu. in. = 3 cu. ft. 1728 cu. in.
Step 2. 1 cu. ft. + 2 cu. ft. = 3 cu. ft.
Step 3. 1728 cu. in. = 1 cu. ft.
Step 4. 3 cu. ft. + 1 cu. ft. = 4 cu. ft.

Explanation. • Add cubic feet in the two measurements for a total of 3 cu. ft. • Add cubic inches and change the total (1,728 cu. in.) to 1 cu. ft. • Add the cubic foot to 3 cu. ft. for a total of 4 cu. ft.

Subtraction

Illustration. A concrete container will hold 4 cu. yd. 10 cu. ft. of concrete. If there are 2 cu. yd. 15 cu. ft. of concrete in the container, how much should be added to fill it?

Solution.   4 cu. yd. 10 cu. ft. = 3 cu. yd. 37 cu. ft.
− 2 cu. yd. 15 cu. ft. = 2 cu. yd. 15 cu. ft.
                            1 cu. yd. 22 cu. ft.

Explanation. • Since the subtrahend (15 cu. ft.) is larger than the minuend (10 cu. ft.), borrow 1 cu. yd. from 4 cu. yd. • Change the borrowed cubic yard to 27 cu. ft. • Add the cubic feet, making a total of 37 cu. ft. • Subtract 2 cu. yd. 15 cu. ft. from 3 cu. yd. 37 cu. ft. for an answer of 1 cu. yd. 22 cu. ft.

Multiplication

Illustration. A *pier* is a pillar which supports an arch on a construction project, figure 30-4. Pier excavation for a building is 4 cu. yd. 12 cu. ft. What is the amount of excavation necessary for the foundation wall if it is 4 times that amount?

Solution.   Step 1. 4 cu. yd. 12 cu. ft. x 4 = 16 cu. yd. 48 cu. ft.
Step 2. 48 cu. ft. ÷ 27 cu. ft. = 1 cu. yd. 21 cu. ft.
Step 3. 16 cu. yd. + 1 cu. yd. 21 cu. ft. = 17 cu. yd. 21 cu. ft.

Explanation. Multiply the size of the pier (4 cu. yd. 12 cu. ft.) by 4 for a total of 16 cu. yd. 48 cu. ft. • Change 48 cu. ft. to 1 cu. yd. 21 cu. ft. by dividing by 27. • Add 16 cu. yd. for a total of 17 cu. yd. 21 cu. ft.

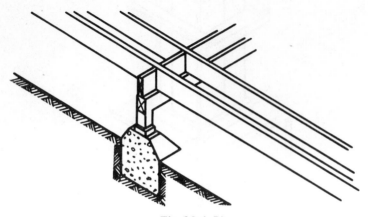

Fig. 30-4 Pier

Division

When dividing cubic units of different denominations, change all the units to the smaller denomination and divide by the given number.

Illustration. If there are 2 cu. yd. 6 cu. ft. of gravel to be equally placed in 6 holes, how much gravel should be placed in each hole?

Solution.   2 cu. yd. 6 cu. ft. ÷ 6
          Step 1. 2 x 27 cu. ft. = 54 cu. ft.
          Step 2. 54 cu. ft. + 6 cu. ft. = 60 cu. ft.
          Step 3. 60 cu. ft. ÷ 6 = 10 cu. ft. for each hole

Explanation. • Change 2 cu. yd. 6 cu. ft. to cubic feet by multiplying 2 by 27 cu. ft. and adding 6 cu. ft. The result is 60 cu. ft. • Divide 60 cu. ft. by 6, which equals 10 cu. ft. to be placed in each hole.

APPLICATION

1. Change cubic feet to cubic inches in each of the following.

    a. 4 cu. ft.   g. 2 1/4 cu. ft.
    b. 9 cu. ft.   h. 5.5 cu. ft.
    c. 8 cu. ft.   i. 6.25 cu. ft.
    d. 20 cu. ft.   j. 2.750 cu. ft.
    e. 86 cu. ft.   k. 5 cu. ft.
    f. 8 1/2 cu. ft.

2. Change cubic inches to cubic feet in each of the following.

    a. 1,728 cu. in.   g. 7,776 cu. in.
    b. 26,882 cu. in.   h. 30,604 cu. in.
    c. 3,456 cu. in.   i. 6,184 cu. in.
    d. 4,320 cu. in.   j. 69,984 cu. in.
    e. 3,888 cu. in.   k. 2,538 cu. in.
    f. 432 cu. in.

3. Change cubic yards to cubic feet in each of the following.

    a. 4 cu. yd.   g. 5 cu. yd.
    b. 85 cu. yd.   h. 76 cu. yd.
    c. 108 cu. yd.   i. 182 cu. yd.
    d. 800 cu. yd.   j. 2,080 cu. yd.
    e. 98.8 cu. yd.   k. 101.12 cu. yd.
    f. 52 1/2 cu. yd.

4. Change cubic feet to cubic yards in each of the following.

    a. 27 cu. ft.   e. 216 cu. ft.   i. 864 cu. ft.
    b. 54 cu. ft.   f. 688 cu. ft.   j. 2,700 cu. ft.
    c. 135 cu. ft.   g. 270 cu. ft.   k. 1,864 cu. ft.
    d. 108 cu. ft.   h. 432 cu. ft.

5. Change cubic yards to cubic inches in each of the following.

  a. 2 cu. yd.   e. 5 cu. yd.   i. 10 1/2 cu. yd.
  b. 4 cu. yd.   f. 7 cu. yd.   j. 12.5 cu. yd.
  c. 1 1/2 cu. yd.   g. 3/4 cu. yd.   k. .9 cu. yd.
  d. .75 cu. yd.   h. 1/8 cu. yd.

6. Change cubic inches to cubic yards in each of the following.

  a. 1,728 cu. in.   e. 26,884 cu. in.   i. 345,600 cu. in.
  b. 3,456 cu. in.   f. 2,727 cu. in.   j. 864 cu. in.
  c. 4,320 cu. in.   g. 34,560 cu. in.   k. 172,800 cu. in.
  d. 17,280 cu. in.   h. 432 cu. in.

7. There are 38,016 cu. in. in the foundation wall shown in figure 30-5. How many cubic feet are there?

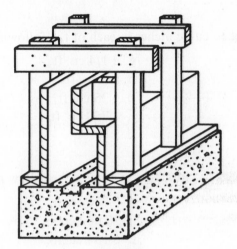

Fig. 30-5

8. If 8 piers contain 10 cu. ft. of concrete apiece, how many cubic feet are required to fill the 8 piers together?

9. Three piers contain the following amounts of concrete: 1 cu. yd. 18 cu. ft., 2 cu. yd. 13 cu. ft., and 23 cu. ft. How much concrete will the 3 piers hold together?

10. There are four sides of a square foundation to be filled equally with concrete from 54 cu. yd. 9 cu. ft. How much should be allowed for each wall?

11. Five cu. yd. 9 cu. ft. of dirt are excavated from one lot and 3 cu. yd. 18 cu. ft. from another lot. What is the difference in the amount of dirt excavated?

Unit 31
Volume of Rectangular Solids and Cubes

OBJECTIVES

After studying this unit, the student should be able to

- find the volume of rectangular solids and cubes.

- determine weight and cost of building materials by volume.

FINDING THE VOLUME OF A RECTANGULAR SOLID

Rectangular solids may be completely solid or "hollow solids". For example, a concrete block is considered a regular solid while the form into which the concrete is poured is considered a hollow solid.

The formula used to determine volume is the same for the two types since both are rectangular; have 90° corners; and have length, width, and height. The formula used to determine volume of a rectangular solid is V = lwh. (V is volume, l is length, w is width, and h is height.)

Illustration. Find the volume in cubic feet of the object in figure 31-1.

Solution. V = lwh
          Step 1. V = 24" x 12" x 6" = 1728 cu. in.
          Step 2. 1728 cu. in. ÷ 1728 cu. in. = 1 cu. ft.

Explanation. • The volume is found by multiplying the length (24") times the width (12") times the height (6"). • The product is 1,728 cu. in., which is changed to cubic feet by dividing by 1,728. The quotient is 1 cu. ft.

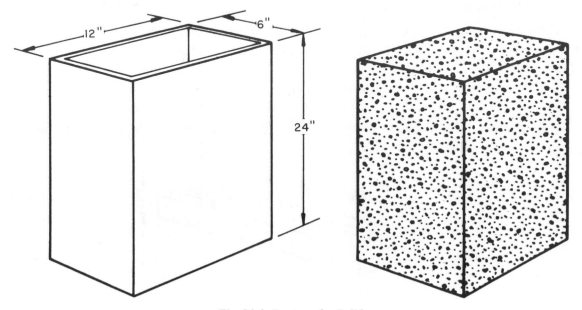

Fig. 31-1 Rectangular Solids

FINDING THE VOLUME OF A CUBE

Determining the volume of a cube is similar to determining the volume of the rectangular solid except that the sides of a cube are equal in linear measure, figure 31-2.

The volume of a cube is equal to the length of a side multiplied by itself three times, $V = S^3$.

Illustration. Find the volume of the cube in figure 31-2. Change cubic inches to cubic feet.

Solution. $V = S^3/1728$
  Step 1. $V = 12'' \times 12'' \times 12''/1728$
  Step 2. $V = 1728$ cu. in./1728 = 1 cu. ft.

Explanation. The sides of the cube measure 12''. The volume is equal to 12'' x 12'' x 12'', or 1,728 cu. in. To change 1,728 cu. in. to cubic feet divide 1,728 cu. in. by 1,728. The quotient is 1 cu. ft.

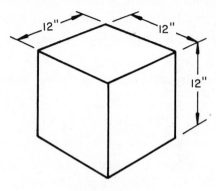

Fig. 31-2 Cube

DETERMINING VOLUME OF CONCRETE

In present day construction, concrete is used on almost every job. Concrete is sold by dealers in cubic yards. The formula used to determine the volume of rectangular solids or cubes can be used to determine the amount of concrete needed for a job.

At times, volume must be determined in cubic feet. The number of cubic feet can be determined by multiplying thickness in feet times width in feet times length in feet.

Illustration. The concrete footing in figure 31-3 is 1' thick, 2' wide, and 10' long. How many cubic feet are in the footing?

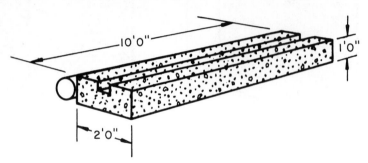

Fig. 31-3 Concrete Footing

Solution.   T′ x W′ x L′
1′ x 2′ x 10′ = 20 cu. ft.

Explanation.   • Multiply the thickness in feet (1′) times the width in feet (2′) times the length in feet (10′), which equals 20 cu. ft.

Since concrete is sold by the cubic yard, the volume should be expressed in cubic yards when possible. This can be done by dividing by 27.

Illustration. How many cubic yards of concrete are needed to pour the wall form in figure 31-4?

Solution.   Cu. yd. = $\dfrac{T' \times W' \times L'}{27}$

$\dfrac{1' \times 4' \times 34'}{27} = \dfrac{136'}{27}$

5 1/27 cu. yd.

Explanation. • The number used in the division is 27 cu. ft. since there are 27 cu. ft. in 1 cu. yd. • Multiply 1′ times 4′ times 34′ and then divide by 27. The answer is 5 1/27 cu. yd.

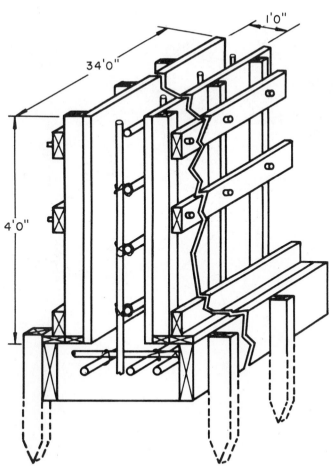

Fig. 31-4 Foundation Footing and Wall

On many jobs, such as the construction of sidewalks and concrete floors, thickness is expressed in inches. When this is the case, 12 should be included in the denominator of the formula so that all dimensions are expressed in feet.

Illustration. How many cubic yards of concrete are needed to pour the sidewalk in figure 31-5?

Solution. $\text{Cu. yd.} = \dfrac{T'' \times W' \times L'}{12 \times 27}$

$$\dfrac{\overset{1}{\cancel{4''}} \times 4' \times \overset{4}{\cancel{108'}}}{\underset{3}{\cancel{12}} \times \underset{1}{\cancel{27}}} = \dfrac{16}{3} = 5\ 1/3 \text{ cu. yd.}$$

Explanation. The thickness of the walk is 4″, which is expressed as 4/12′. • The 27 is used in division so that the measurement is expressed in cubic yards. • Multiply 4 x 4 x 108 and divide by 12 x 27. The result is 5 1/3 cu. yd.

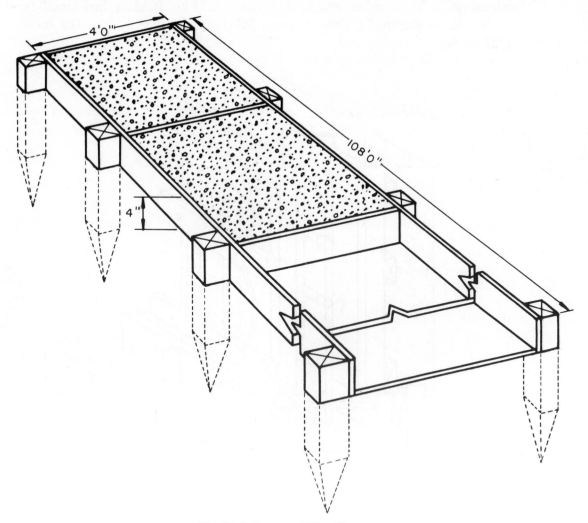

Fig. 31-5 Concrete Sidewalk

DETERMINING AMOUNTS OF AGGREGATES IN A CONCRETE MIXTURE

Aggregates are particles of different substances which make up a compound or mass. Concrete is a mixture of cement, sand, and gravel. It may be classified as rich, medium, ordinary, or lean. Once the proportion of materials in 1 cu. ft. of concrete is known, the proportion of materials in the entire mixture can be determined. For example, a medium mixture of concrete, with a proportion of 1:2 1/2:5 (one part cement to 2 1/2 parts sand to 5 parts gravel) could contain 0.048 barrels, or bbl., of cement, 0.0170 cu. yd. sand, and 0.0341 cu. yd. gravel. These measurements are constants based on the given proportion. *Note:* In construction, a barrel contains 31 1/2 gallons.

To determine the amount of each aggregate in the mixture, first determine the number of cubic feet of volume. Multiply the total volume in cubic feet by the volume constant for each material.

Illustration. Find the amount of cement, sand, and gravel needed to pour a carport floor 4″ thick, 21′ wide, and 22′ long.

Solution.   (4″ x 21′ x 22′) ÷ 12 = 154 cu. ft.
Cement = 154 x 0.048 bbl. = 7.392 bbl.
Sand = 154 x 0.0170 cu. yd. = 2.6180 cu. yd.
Gravel = 154 x 0.0341 cu. yd. = 5.2514 cu. yd.

Explanation. • Multiply all measurements (4″ x 21′ x 22′) and divide by 12 to find the volume of the rectangular solid. • Multiply the total (154 cu. ft.) by each proportion for the total amount of cement, sand, and gravel needed to pour the carport.

Note: Particles of sand, cement, and gravel expand as they are mixed with water.

The same procedure can be used to determine the proportions of cement and sand in a mixture.

Illustration. Find the amount of cement and sand needed for a floor 1″ thick, 18′ wide, and 24′ long, with a 1:2 mixture.

Solution.   T″ x W′ x L′ = total cu. ft.
(1″ x 18′ x 24′) ÷ 12 = 36 cu. ft.
Cement = 36 x 0.1234 = 4.4424 bbl.
Sand = 36 x 0.0344 = 1.2384 cu. yd.

Explanation. • Find the volume of the floor by multiplying 1″ x 18′ x 24′ and dividing by 12. The result is 36 cu. ft. • Multiplying 36 by 0.1234 gives the amount of cement (4.4424 bbl.). Multiplying 36 by 0.0344 gives the amount of sand (1.2384 cu. yd.).

DETERMINING WEIGHT AND COST OF MATERIALS

The cost of materials can be determined by volume measurement after the weight of the material is known.

Illustration. A piece of brass which weighs .30 lb. per cu. in. is purchased. It is 3/4″ thick, 2″ wide, and 44″ long. What is the total weight of the bar? What is its cost if it sells for 75¢ per pound?

Solution.   Step 1. V = T″ x W″ x L″

$$V = 3/4″ \text{ x } 2″ \text{ x } 44″ = \frac{\overset{}{3}″}{\underset{1}{\cancel{4}}} \text{ x } 2″ \text{ x } \overset{11}{\cancel{44}}″ = 66 \text{ cu. in.}$$

V = 66 cu. in.

Step 2. Wt = 66 x .30 = 19.80 lb.

Step 3. Cost = 19.80 x .75 = 14.8500

Cost = $14.85

Explanation. • The product of T″ x W″ x L″ gives the volume in cubic inches (66 cu. in.). • Multiply .30, the weight of brass per cubic inch (.30), by 66 cu. in. This gives the weight of the brass (19.80 lb.). • Multiply 19.80 by .75 for the cost of the brass ($14.85).

APPLICATION

1. How many cubic inches are in a box which measures 10″ by 22″ by 80″?

2. How many cubic inches are in a box measuring 12 1/2″ by 14″ by 22 3/4″?

3. How many cubic feet are in a box 3′ by 4′ by 6′?

4. How many cubic feet are in a box 2.8′ by 3.5′ by 7′?

5. How many cubic yards of dirt are excavated if the plot of dirt measures 36′ by 60′ by 6′?

6. How many cubic yards of dirt are excavated if the plot has sides measuring 6′?

7. If the volume of a rectangular solid is given in cubic inches, how is the volume changed to cubic feet?

8. If the volume of a cube is given in cubic inches, how is it changed to cubic feet?

9. If the volume of a rectangular solid is given in cubic feet, how is it changed to cubic yards?

10. How many cubic yards of concrete are required to fill 12 concrete forms like the one in figure 31-6?

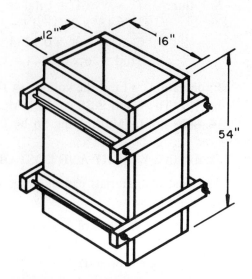

Fig. 31-6

11. How many cubic yards of concrete are required to pour a sidewalk 4" thick, 4' wide, and 72' long?

12. How many cubic yards of concrete are required to pour a concrete slab 4" x 27' x 62'?

13. What is the cost of concrete to pour a driveway 4" x 10' x 96' if the concrete sells for $16.25 per cubic yard?

14. If a wooden beam is 10" square and 16' long, how many cubic feet does the beam measure?

15. If the measurement of a cube is 1' on all sides, how many cubic feet does the cube measure?

16. How many cubic feet of concrete are required to pour a beam measuring 12" x 12' x 20'?

Unit 32
Volume of Cylinders, Cones, Pyramids and Spheres

OBJECTIVES

After studying this unit, the student should be able to

- find the volume of a cylinder, pyramid, and sphere.

- find the volume of a cone and a frustum.

FINDING THE VOLUME OF A CYLINDER

Review the definition of a cylinder found in Unit 29. To find the volume of a cylinder, use one of two formulas: $V = \pi r^2 h$ or $V = 1/4\pi d^2 h$. (When either the diameter or radius is given, the other can always be determined.)

Illustration 1. A *dry cell* provides electrical current and consists of three elements: a zinc cylinder, a paste electrolyte, and a carbon rod. The dry cell in figure 32-1 is in a cylindrical shape. The diameter measures 3″ and the height is 8″. What is its volume? Use the formula $V = \pi r^2 h$.

Solution. $V = \pi r^2 h$
$V = 3.1416 \times 2.25'' \times 8''$
$V = 56.5488$ cu. in.

Explanation. • The diameter (3″) is used to determine the radius (1.5″), which is squared (1.5″ x 1.5″ = 2.25″). • Multiply 3.1416 (π) x 2.25″ x 8″, which equals 56.5488, the number of cubic inches in the battery.

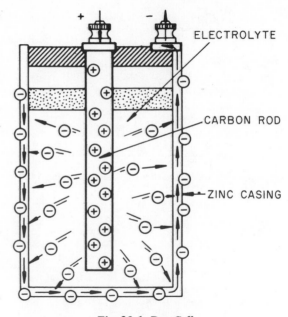

ELECTROLYTE

CARBON ROD

ZINC CASING

Fig. 32-1 Dry Cell

Illustration 2. Find the volume of the dry cell in figure 32-1 using the second formula given, $V = 1/4\pi d^2 h$.

Solution. $V = 1/4\pi d^2 h$

Step 1. $\dfrac{1}{\cancel{4}} \times \overset{.7854}{\cancel{3.1416}} \times 9 \times 8$
1

Step 2. $V = 1 \times .7854 \times 9 \times 8 = 56.5488$

Explanation. • Find the value of the diameter squared (6^2) by multiplying 3 by 3. • Then multiply 1/4 x 3.1416 (pi) x 9 x 8. The product is 56.5488 cubic inches, the same answer found in Illustration 1.

FINDING THE AREA OF A RIGHT PYRAMID

A pyramid may be either a geometrical shape or a 3-dimensional object. As an object, it has walls in the form of triangles that meet in a point at the top. The point at the top of the pyramid is called the *vertex* or *apex*. The base of a pyramid may be square, rectangular, or have several sides. Figure 32-2 shows a pyramid with a square bottom.

To find the volume of a pyramid, use one of two formulas: V = 1/3b (area of base) h or V = bh/3. That is, multiply 1/3 by the area of the base by the height.

Illustration. Find the volume of the pyramid in figure 32-2.

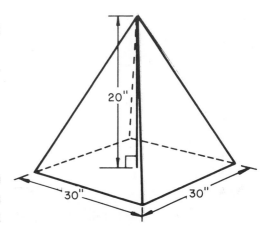

Solution. V = 1/3 x area of base x height
Step 1. b = 30″ x 30″ = 900 sq. in.

Step 2. $V = \dfrac{1}{\cancel{3}} \times \cancel{900}^{\,300} \times 20 =$

6000/1
V = 6000 cu. in.

Explanation. • The area of the base is 900 sq. in. (30 by 30 = 900) and the height is 20″. • Following the formula, 1/3 x 900 x 20 = 6000 cubic inches.

Fig. 32-2 Pyramid with Square Bottom

FINDING THE VOLUME OF A CONE

A *cone* is an object or a geometrical shape with a circular base and sides which taper evenly to a point.

To determine the volume of a cone, multiply 1/3 by the area of the base by the height. The formula is V = 1/3πr²h or V = πr²h/3.

Illustration. Find the volume of the cone in figure 32-3.

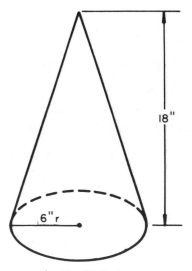

Solution. V = 1/3πr²h
Step 1. $V = \dfrac{1}{\cancel{3}} \times 3.1416 \times$
1

$36 \times \cancel{18}^{\,6}$

Step 2. V = 1 x 3.1416 x 36 x 6 = 678.5856/1 cu. in.
V = 678.59 cu. in.

Explanation. Following the formula, multiply 1/3 times pi (3.1416) times the radius squared (36 sq. in.) times the height (18″). The product is the volume of the cone, 678.59 cu. in.

Fig. 32-3 Cone

FINDING THE VOLUME OF A FRUSTUM

A *frustum* is formed by cutting off the top of a cone-shaped solid with a plane parallel to the base, figure 32-4. A funnel is an example of a frustum cone.

The volume of a frustum may be found by using the formula $V = 1/3\pi h (R^2 + r^2 + Rr)$.

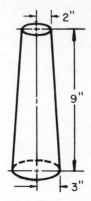

Fig. 32-4 Frustum

Illustration. Find the volume of the frustum in figure 32-4 with the given dimensions. In the figure, the height (h) is 9″, the base radius (r) is 3″, and the top radius (r) is 2″.

Solution. $V = 1/3\pi h (R^2 + r^2 + Rr)$

Step 1. $V = \dfrac{1}{\cancel{3}} \times 3.1416 \times \cancel{9}^{3} (3^2 + 2^2 + 3 \times 2)$

Step 2. $V = 1 \times 3.1416 \times 3 (9 + 4 + 6)$

Step 3. $V = 9.4248 \times 19$

$V = 179.0712$ cu. in.

Explanation. • Multiply 1/3 times pi (3.1416) times the height (9″). • Multiply this product by R^2 (9 sq. in.) plus r^2 (4 sq. in.) plus Rr (6 sq. in.). The answer is 179.0712 cu. in., the volume of the frustum.

FINDING THE VOLUME OF A SPHERE

A *sphere* is a round object, such as a ball, with all points of the surface equidistant from the center, figure 32-5. The volume of a sphere is found by multiplying 4 times pi times the cube of the radius, and dividing the product by 3.

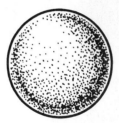

Fig. 32-5 Sphere

Illustration. Find the volume of the sphere in figure 32-5.

Solution. $V = \dfrac{4\pi r^3}{3}$

  Step 1. $V = (4 \times 3.1416 \times 2^3)/3$
  Step 2. $V = (4 \times 3.1416 \times 8)/3$
  Step 3. $V = 100.5312/3$
       $V = 33.5104$ cu. in.

Explanation. The volume of the sphere is found by multiplying 4 x 3.1416 times 8. The product (100.5312) is divided by 3. The quotient is 33.5104 cu. in.

APPLICATION

1. What is the volume of a cylinder if the radius of its base is 4″ and the height is 8″?

2. What is the volume of a cylinder if the diameter of its base is 12′ and the height is 18′?

3. What is the volume of a cylinder if the diameter of its base is 4′ 4″ and the height is 8′ 8″?

4. What is the volume of a cylinder if the radius of its base is 3.2′ and the height is 9.3′?

5. Which container holds more and how much more does it hold: one 4″ in diameter and 8″ high or one 2″ in radius and 8″ high?

6. One cubic foot equals 7 1/2 gallons of liquid measure. How many gallons will a drum with a 20′ diameter and 44′ height hold?

7. How many cubic feet of water will a water tank hold if its radius is 24″ and height 60″?

8. A grain tank 22′ deep has an inside diameter of 15′. How many cubic feet of grain will it hold?

9. Find the volume of a pryamid with one side of its square base measuring 30″ and the height measuring 33″.

10. Find the volume of a pyramid with a base 30″ long and 15″ wide and a height of 45″.

11. What is the volume of a pyramid with a base of 12′ by 16.5′ and a height of 25′?

12. How many cubic feet of space are inside the pyramid in figure 32-6 if its square base measures 14′ and the height is 14′?

Fig. 32-6

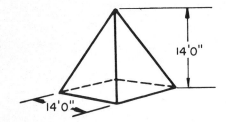

14'0"   14'0"

13. Find the volume of a right cone whose base is 10″ in diameter and 8″ high.

14. Find the volume of a cone which is 9 1/2″ high and has a base with a 3″ radius.

15. Find the volume of a frustum which is 10″ high. The base radius is 4″ and the top radius is 2″.

16. Find the volume of a frustum with a height of 14″. The base radius is 3″ and the top radius is 1 1/2″.

17. Find the volume of the following spheres as indicated.

   a. 22″ sphere
   b. 5″ sphere
   c. 3″ sphere
   d. sphere with a 2.1″ radius
   e. sphere with a 3/4″ diameter
   f. sphere with a 3.2″ radius

CAREER PROFILE: THE STAIR BUILDER

Job Description

The stair builder is a specialized carpenter who constructs a series of uniform steps leading from one level of a structure to another. These persons must select materials for, lay out, and cut out stringers, treads, and risers. They also read blueprints to determine the size of the stairwell and headroom. Jobs are usually indoors with few hazards. Many stair builders work in mills. The stair builder who works on the job site begins after the rough construction is finished.

Qualifications

Courses in general woodwork as well as a knowledge of measuring and cutting tools are necessary for the stair builder. Competency in mathematics and knowledge of portable power tools and materials is important.

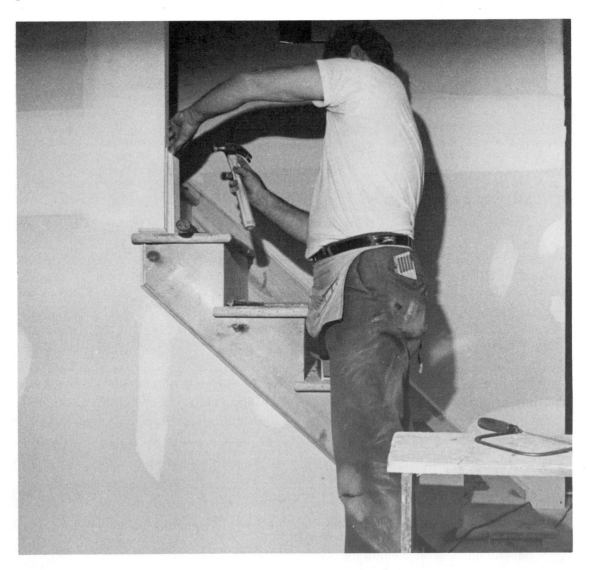

Section 5 The Metric System of Measurement

Unit 33   Using the Metric System: Linear, Volume, and Weight Measurements

OBJECTIVES

After studying this unit, the student should be able to

- change metric units of one denomination to units of another denomination.

As carpenters deal more with the metric system, they may find the need to change units within the metric system. Figure 33-1 shows commonly used metric equivalents.

LINEAR MEASURE

10 millimeters (mm) = 1 centimeter (cm)
10 centimeters or 100 millimeters (mm) = 1 decimeter (dm)
10 decimeters (dm) or 100 centimeters (cm) = 1 meter (m)
10 meters (m) or 1000 centimeters (cm) = 1 dekameter (dam)
1 centimeter (cm) = 10 millimeters (mm)
1 meter (m) = 100 centimeter (cm)
1 meter (m) = 1 000 millimeters (mm)

WEIGHT MEASURE

10 milligrams (mg) = 1 centigram (cg)
10 centigrams = 1 decigram (dg)
10 decigrams = 1 gram (g)
10 grams = 1 dekagram (dag)

VOLUME MEASURE

10 milliliters = 1 centiliter (cl)
10 centiliters = 1 deciliter (dl)
10 deciliters = 1 liter (l)
10 liters (l) = 1 dekaliter (dal)

Fig. 33-1 Metric Equivalents

To change metric units of one denomination to units of another denomination, simply multiply or divide by the proper number, as in changing units in the English System. Refer to figure 33-1 when necessary. The following illustrations show some examples of conversion. After studying the table and examples, the student should be able to freely change metric units.

CHANGING CENTIMETERS TO MILLIMETERS

To change centimeters to millimeters, multiply the number of centimeters by 10.

Illustration. Change 5 cm to millimeters. Refer to figure 33-2.

Fig. 33-2

Solution. 10 mm x 5 = 50 mm

Explanation. Multiply 10 by 5 cm to equal 50 mm.

CHANGING METERS TO CENTIMETERS

To change meters to centimeters, multiply the number of meters by 100 cm.

Illustration. Change 6 m to centimeters.

Solution. 100 cm x 6 = 600 cm

Explanation. Multiply 100 cm x 6. The product is 600 cm. There are 600 cm in 6 m or 100 cm in each of the 6 m.

CHANGING MILLIMETERS TO CENTIMETERS

To change millimeters to centimeters, divide the number of millimeters by 10.

Illustration. Change 34 mm to centimeters.

Solution. 34 ÷ 10 cm = 3.4 cm

Explanation. Divide 34 by 10 since there is 1 cm for every 10 mm. The result is 3.4 cm.

CHANGING LITERS TO MILLILITERS

To change liters to milliliters, multiply the number of liters by 1 000.

Illustration. Express the measurement shown on the container in figure 33-3 in milliliters.

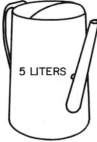

5 LITERS

Fig. 33-3

Solution. 5 x 1 000 ml = 5 000 ml

Explanation. Since there are 5 liters, the results are equal to 5 times 1 000, the number of milliliters in 1 liter. The answer is 5 000 ml.

CHANGING SQUARE METERS TO SQUARE CENTIMETERS

Remember that square area measurement is the length of the side of an object or figure times itself. Figure 33-4, page 200, gives square area equivalents in the metric system with their abbreviations.

100 square millimeters (mm^2) = 1 square centimeter (cm^2)
100 square centimeters (cm^2) = 1 square decimeter (dm^2)
100 square decimeter (dm^2) = 1 square meter (m^2)

Fig. 33-4 Square Area Equivalents

To change square meters to square centimeters, multiply the number of square meters by 100 twice.

Illustration. Change 6 m^2 to square centimeters.

Solution.   6 x 100 x 100 = 60 000 cm^2

Explanation. • Change 6m^2 to square decimeters by multiplying 6 by 100. • Change this product (600) to square centimeters by again multiplying by 100. The answer is 60 000 cm^2.

CHANGING CUBIC MILLIMETERS TO CUBIC CENTIMETERS

Three common measures of volume in the metric system are cubic centimeters, cubic decimeters, and cubic meters. Figure 33-5 shows volume equivalents for cubic measurements.

1 000 cubic millimeters (mm^3) = 1 cubic centimeter (cm^3)
1 000 cubic centimeters (cm^3) = 1 cubic decimeter (dm^3)
1 000 cubic decimeters (dm^3) = 1 cubic meter (m^3)

Fig. 33-5 Cubic Equivalents

Cubic measurements may be changed from one denomination to another denomination by division or multiplication.

Illustration. Change 4 800 mm^3 to cubic centimeters.

Solution.   4 800 ÷ 1 000 = 4.8 cm^3

Explanation. There are 1 000 mm^3 in 1 cm^3. Divide 4 800 by 1 000. The quotient is 4.8 cm^3.

Note: The exponent (^{3}) is a figure used to represent the word "cubic".

APPLICATION

1. Convert the following measurements as indicated.

   a. 8 cm to millimeters
   b. 12 dm to centimeters
   c. 6 m to decimeters
   d. 31 m to decimeters
   e. 10.5 cm to millimeters
   f. 8.4 dm to centimeters

   g. 20.8 m to decimeters
   h. 8.8 m to centimeters
   i. 45 cm to millimeters
   j. 55 dm to centimeters
   k. 18.55 m to centimeters

2. Convert the following measurements of weight as indicated.

   a. 7 cg to milligrams
   b. 18 dg to centigrams
   c. 12 g to decigrams
   d. 25 g to centigrams
   e. 30 mg to centigrams

   f. 23 cg to decigrams
   g. 6.75 cg to milligrams
   h. 8.6 g to milligrams
   i. 8.8 dg to grams
   j. 85 mg to grams

3. Convert the following measurements of volume as indicated.

   a. 7 cm^3 to cubic millimeters   d. 6 800 dm^3 to cubic meters
   b. 8.5 dm^3 to cubic centimeters e. 38 000 dm^3 to cubic meters
   c. 300 mm^3 to cubic centimeters

4. Convert the following square measurements as indicated.

   a. 9 cm^2 to square millimeters   e. 100 mm^2 to square centimeters
   b. 85 dm^2 to square centimeters f. 13 dm^2 to square meters
   c. 40 m^2 to square decimeters   g. 20 m^2 to square millimeters
   d. 6.5 m^2 to square centimeters

5. The length of a line is 6 cm. How many millimeters are there in the line?

6. The perimeter of a rectangle is 9 m. Change 9 m to centimeters.

7. The circumference of a circle is 36 mm. What is the circumference expressed in centimeters?

8. The layout of a room shows that the room is 12 m wide and 14 m long. What are the measurements in centimeters?

Unit 34   Using the Metric System:
Square and Cubic Measurement

OBJECTIVES

After studying this unit, the student should be able to

- change square unit measurements of the metric system from one denomination to another.

- change cubic unit measurements of the metric system from one denomination to another.

SQUARE MEASUREMENT IN THE METRIC SYSTEM

Square unit measurements correspond with linear measurements. For example, a square, each side of which measures 1m, is called a square meter. When the measurement is increased to 2 m on each side, the area is not 2 sq. m; it is 2 m times 2 m which is 4 m. Figure 34-1 gives common square equivalents in the metric system.

100 square millimeters = 1 square centimeter (cm^2)

100 square centimeters = 1 square decimeter (dm^2)

100 square decimeters = 1 square meter (m^2)

Fig. 34-1 Metric Equivalents (Square Measurement)

CHANGING SQUARE UNITS

After studying figure 34-1 and the following example, the carpentry student should be able to change all basic square unit measurements in the metric system.

Illustration. How many square millimeters does the metal sheet in figure 34-2 contain?

Solution.   Step 1. 14 cm x 18 cm = 252 cm²
                  Step 2. 252 x 100 mm² = 25 200 mm²

Explanation. • The square area of the metal is found by multiplying the length times the width (14 cm x 18 cm). • Change the product (252 cm²) to square millimeters. Since there are 100 mm² in 1 sq. cm, multiply 252 by 100 which equals 25 200 mm².

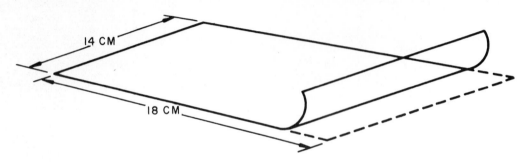

14 CM

18 CM

Fig. 34-2 Sheet Metal

CUBIC MEASUREMENT IN THE METRIC SYSTEM

The volume of a solid is obtained by cubic measurement. Just as in the English System of measurement, there are three dimensions in cubic measurement in the metric system: length, width, and thickness, figure 34-3. The only difference in the two systems is the manner of expression.

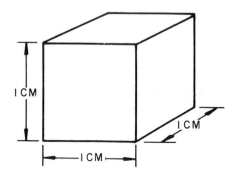

Fig. 34-3 Cube with Volume of 1 cm^3

The content of a cube is expressed in volume. The volume is equal to the product of the length times the width times the thickness. The volume is expressed in cubic measurement. For example, to find the volume of the cube in figure 34-3, multiply 1 cm x 1 cm x 1 cm. The volume is 1 cm^3.

If an object has length, width, and height which do not measure the same, the volume is still expressed in cubic measurement though the object itself is not a cube.

1 000 cubic millimeters = 1 cubic centimeter (1 cm^3) or 1 cube
1 000 cubic centimeters = 1 cubic decimeter (1 dm^3) or 1 liter
1 000 cubic decimeters = 1 cubic meter (1 m^3) or 1 kiloliter (kl)

Fig. 34-4 Metric Equivalents (Cubic Measurement)

CHANGING CUBIC UNITS

Just as in square unit measurement, changing cubic units should be easy for the carpentry student after studying the following example and figure 34-4. Simple multiplication or division is used.

Illustration. A box measures 2 200 cubic millimeters. Express the measurement in cubic centimeters.

Solution. 2 200 ÷ 1 000 = 2 cubic centimeters, 200 cubic millimeters

Explanation. To change the measurement, divide 2 200 cubic millimeters by the proper number as shown in figure 34-4 (1 000). The answer is 2 cubic centimeters, 200 cubic millimeters.

APPLICATION

1. A ceiling light contains a piece of glass 32 cm by 24 cm. How many square millimeters of glass does it contain?

2. A piece of sheet metal measures 16 cm by 8 cm. What is its area in square inches?

3. How many square centimeters are there in an area 25 cm long and 18 cm wide?

4. A piece of sheet metal is 1 m wide and 100 cm long. How many square centimeters are there in the piece of sheet metal?

5. A piece of metal is 10 cm wide and 2 dm long. How many square decimeters are there in the piece of metal?

6. A sign board is 3 m by 4 m. It costs 2¢ per square centimeter to make the sign board. What is the total cost?

7. Find the area of a rectangle 17 m wide and 21 m long.

8. Find the perimeter of a rectangle which is 37 cm wide and 42 cm long.

9. What is the area of a triangle if its height is 34 cm and its base is 20 cm?

10. Find the area of a square with sides measuring 14 mm.

11. A board measures 3 dm^3. What is the measurement in cubic centimeters?

12. A carpenter has a box which measures 2 cm^3. What is the measurement in cubic millimeters?

CAREER PROFILE: THE CABINETMAKER

Job Description

Cabinetmakers construct kitchen and bathroom cabinets and closets. An experienced cabinetmaker may also construct various pieces of furniture. Cabinetmakers may be employed in specialized plants or mills, usually located in large urban areas. Cabinets built at these plants are then shipped to other areas of the country. Cabinetmakers may also work on job sites or remain self-employed. The cabinetmaker is a true craftsman who must possess very specialized abilities.

Qualifications

Cabinetmakers must have a working knowledge of different wood types and methods of shaping and cutting wood. They must be able to read blueprints, make freehand sketches, and measure and estimate materials. Many times, the cabinetmaker may use his creative ability to design furniture. Since the cabinetmaker is constantly working with power tools, a knowledge of these tools and safety practices is essential.

Installing the Cabinets on the Job in a Typical House

Section 6
Square Roots

Unit 35
Squaring and Square Roots

OBJECTIVES

After studying this unit, the student should be able to

- square numbers.
- determine square roots.

SQUARING A NUMBER

The product obtained by multiplying a number by itself is the *square* of that number. For example, the square of 6 is 36, the product obtained when 6 is multiplied by itself (6 x 6 = 36).

The 6 is a factor of 36; when 6 is squared, it is written 6^2. The small 2 which is written a little above and to the right of the number shows that the number is squared. It is the exponent.

The area of a square can be found by finding the square of a given side.

Illustration. Find the square area of the room whose layout is shown in figure 35-1.

Solution.   12′ x 12′ = 144 sq. ft.
                12^2 = 144 sq. ft.

Explanation. • Multiply the number (12) by itself (12 x 12 or 12^2). • The numbers represent the length and width of the room. The area of the room is 144 sq. ft.

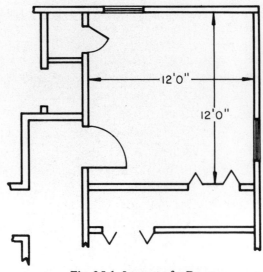

Fig. 35-1 Layout of a Room

Some formulas require the squaring of a number before the problem can be solved. For example, the formula used to determine the area of a circle is $A = \pi r^2$. Before π is multiplied by r^2, the number represented by r must be squared, or multiplied by itself.

Illustration. In figure 35-2, $A = \pi r^2$. What number does r^2 represent?

Solution. $r^2 = 5^2{}''$

$5^2{}'' = 25$ sq. in. or 5″ x 5″ = 25 sq. in.

Explanation. The r^2 is $5^2{}'' = 25$ sq. in., or 5″ x 5″ = 25 sq. in.

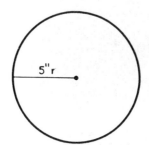

Fig. 35-2 Circle with 5″ Radius

DETERMINING SQUARE ROOTS

The square root of a number is one of two equal factors of the number. If a number contains 2 equal whole number factors, it is a perfect square. For example, 36 is a perfect square because the two factors are 6 and 6 (whole numbers). The square root of 36 is 6 because 6 x 6 equals 36. The square root symbol, $\sqrt{}$, is called a *radical sign*.

Illustration. Find the length and width of a square room with an area of 144 sq. ft.

Solution.

$$
\begin{array}{r}
1\;\;\;2' \\
1\sqrt{1'44}\text{ sq. ft.} \\
\underline{1} \\
22\overline{)44} \\
\underline{44}
\end{array}
$$

Explanation. • Write the number 144 under the square root symbol and divide it into pairs of figures beginning at the right end of the number (1′44). • There are two groups formed and the quotient has 2 digits. The first group to consider (actually, it is only one number) is the figure 1. The largest perfect square contained in 1 is 1; its square root is 1. Write 1 in the quotient. Subtract the square of 1 (1) from the partial dividend (1). There is no remainder. • Bring down the next pair of figures (44) which is the next partial dividend to consider. • The trial divisor is found by multiplying the partial quotient (1) by 2: 2 x 1 = 2. • The trial divisor is in the 10s column, so the trial divisor is between 20 and 29. As a trial, 20 goes into 44 about 2 times. • Annex 2 to 2 which equals 22. The 22 goes into 44 exactly 2 times. • Write 2 in the quotient position, and multiply 2 times 22 which equals 44. • Subtract 44 from 44. • There is no remainder, so the answer is 12. • To prove the answer, multiply the quotient by itself (12 x 12). • The product is 144, which proves that 12 is the square root of 144 sq. ft.

In this example, 144 is known as a *perfect square* because the square root (12) is a whole number. Numbers which have remainders in the square roots are known as *imperfect squares*. Most numbers do have remainders in their square roots. The procedure is basically the same as the procedure for finding the square root of a perfect square. The difference in procedures involves annexing zeros in pairs to the number.

Illustration. Find the square root of 810. Carry out the answer to one decimal place.

Solution.

$$
\begin{array}{r}
2\ 8.\ 4 \\
2\sqrt{8'10.00} \\
\underline{4} \\
48\ )\overline{4\ 10} \\
\underline{3\ 84} \\
564\ )\ \overline{26'00} \\
\underline{22'56} \\
3'44
\end{array}
$$

Proof. 28.4 x 28.4 = 810.00

Explanation. • Write the number 810 under the square root symbol. • Place a decimal point and annex two zeros. • Starting at the right, group the numbers in pairs. After grouping, the number is 8'10.00. • There are three groups; therefore, the quotient has three digits. • Begin division with the first group (8). The largest perfect square contained in 8 is 4, the square root of which is 2. • Write 2 in the quotient. • Subtract 2^2 (4) which is from the partial dividend (8). This leaves 4. • Bring down the next pair (10) to get the next partial dividend (410). • Multiply the quotient 2 by 2, which equals 4, to find the trial divisor. The trial divisor is between 40 and 49. As a trial factor, 40 goes into 410 about 8 times. • Use 48 as the divisor and write 8 in the quotient. • Multiply 8 by 48, which equals 384 with 26 remaining. • Place a decimal point in the quotient and bring down the next pair, making the number 2600. • Multiply the quotient 28 by 2 which equals 56. The number 56, the trial divisor, goes into the number about 4 times. • Write the 4 in the quotient and multiply 4 by 564. This equals 2256. • Subtract 2256 from 2600. The remainder is 344. The answer carried out one decimal place is 28.4, with 344 remaining. To prove the answer, 28.4 is multiplied by itself and the remainder is added. The answer is 810.00, the original number.

Except for one basic change, the procedure for finding the square root of a mixed number is the same. Express the fraction in decimal form and proceed as usual.

APPLICATION

1. Square each of the following numbers.

a. 5	d. 11	g. 13	j. 33	m. 125
b. 9	e. 12	h. 14	k. 7	n. 300
c. 10	f. 6	i. 22	l. 52	o. .2

2. Find the areas of squares whose sides have the following measurements.

    a. 14'   d. 25"   g. 6 1/4'   j. 44"   m. 7/8 yd.
    b. 16'   e. 30"   h. 10 2/5'   k. 1/2 yd.   n. 1/3 yd.
    c. 22"   f. 3 1/2'   i. 20"   l. 3/4 yd.   o. 4/5 yd.

3. Find the square root of the following numbers. If there is a remainder, carry it out to one decimal place.

    a. 25 sq. ft.   d. 144 sq. ft.   g. 800 sq. ft.   j. 51.84 sq. ft.
    b. 225 sq. ft.   e. 676 sq. in.   h. 731 sq. ft.   k. 200 sq. ft.
    c. 576 sq. ft.   f. 5184 sq. in.   i. 741 sq. ft.   l. 2500 sq. ft.

4. Find the square root of the following fractions. Carry out the answer to three decimal places.

    a. 1/3   c. 3/8   e. 1/2   g. 5/8
    b. 3/16   d. 7/16   f. 9/16   h. 11/16

5. A certain room has 4 equal sides with a total of 484 sq. ft. Find the length of one of its sides.

6. A playroom has four equal sides. If the area measures a total of 1625 sq. ft., what is the length of one side? Carry out the answer to 2 decimal places.

Unit 36
Hypotenuse

OBJECTIVES

After studying this unit, the student should be able to

- determine the length of the hypotenuse of any given right angle.

The *hypotenuse* is that side of a right triangle which is located opposite the right angle. The hypotenuse is always the longest side of a triangle. Many problems in carpentry involve finding the length of the hypotenuse, such as squaring corners of a foundation or determining the length of a common rafter or brace.

The *Pythagorean theorem* states that the hypotenuse is equal to the square root of the sum of the squares of the other two sides: $h = \sqrt{a^2 + b^2}$. In the formula, h is the hypotenuse, b is the base, and a is the altitude, figure 36-1.

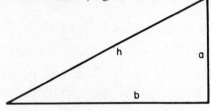

Fig. 36-1 Right Triangle

PRACTICAL APPLICATION

Squaring Foundation Corners

An equivalent of the formula $h = \sqrt{a^2 + b^2}$, known as the 6-8-10 process, is used by carpenters to square corners of foundations. The 6-8-10 process is known as such because the hypotenuse of the triangle formed is equal to the square root of the sum of 6' and 8' squared, figure 36-2.

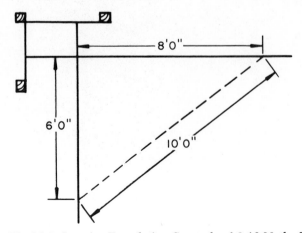

Fig. 36-2 Squaring Foundation Corner by 6-8-10 Method

Illustration. Prove that the square root of the sum of the squares of 6′ and 8′ is equal to 10′.

Solution. $h = \sqrt{a^2 + b^2}$

Step 1. $h = \sqrt{6^2 + 8^2} = \sqrt{36 + 64}$

Step 2.

$$
\begin{array}{r}
1\ \ 0' \\
1\sqrt{1'00} \\
\underline{1\ \ \ } \\
00
\end{array}
$$

Check. $10^2 = 100$ or $10 \times 10 = 100$

Explanation. • Apply the formula; 6^2 equals 36 and 8^2 equals 64. The sum of 36 and 64 is 100. • The square root of 100 is 10′. Prove this fact by multiplying 10 x 10, which equals 100.

The corners of a foundation can be squared by determining the diagonal lengths of the room which are the hypotenuse of the two triangles formed by the diagonals. To form a right triangle, the corners must measure 90°.

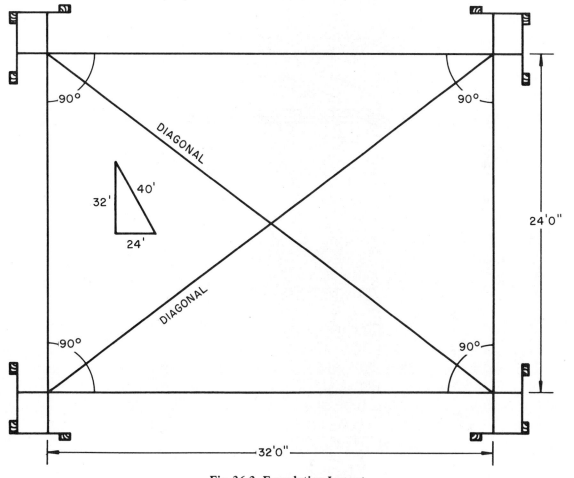

Fig. 36-3 Foundation Layout

Illustration. The room in figure 36-3, page 211, has one side measuring 24′ and the other side measuring 32′. What is the length of the diagonal?

Solution. $h = \sqrt{a^2 + b^2}$

Step 1. $h = \sqrt{24^2 + 32^2} = \sqrt{576 \text{ sq. ft.} + 1024 \text{ sq. ft.}}$

Step 2.

$$
\begin{array}{r}
4\ \ 0' \\
4\sqrt{1600 \text{ sq. ft.}} \\
\underline{16\ \ \ \ } \\
00
\end{array}
$$

Proof. 40′ x 40′ = 1600 sq. ft.

Explanation. The length of the diagonal is the length of the hypotenuse. • Apply the formula $h = \sqrt{24^2 + 32^2}$. • The sum of the square of the two numbers is 1600. • The square root of 1600 is 40′. • To prove the answer, multiply 40 by 40. The product is the original number, 1600.

Common Rafters

Part of the length of a common rafter forms the hypotenuse of a right triangle. To find the length of the common rafter from the building line to the center of the ridge, apply the formula $h = \sqrt{a^2 + b^2}$.

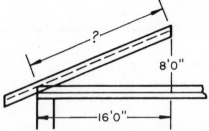

Fig. 36-4 Length of a Common Rafter

Illustration. The rise of the rafter measures 8′ and the run, 16′. Apply the formula $h = \sqrt{a^2 + b^2}$ to find the length of the rafter from the building line to the center of the ridge along the measuring line.

Solution. $h = \sqrt{a^2 + b^2}$ or $h = \sqrt{\text{rise}^2 + \text{run}^2}$

Step 1. $h = \sqrt{8^2 + 16^2}$

Step 2. $h = \sqrt{64 \text{ sq. ft.} + 256 \text{ sq. ft.}}$

Step 3.

$$
\begin{array}{r}
17\ .\ 8\ \ 8' \\
\sqrt{320.00\ 00}
\end{array}
$$

Step 4.

$$
\begin{array}{r}
1' \\
27\overline{)\,220} \\
\underline{189\ \ \ \ } \\
348\overline{)\,3100} \\
\underline{2784\ \ \ } \\
3568\overline{)\ \ 31600} \\
\underline{28544\ } \\
3056
\end{array}
$$

Explanation. • The rise (8′) squared equals 64 and the run (16′) squared equals 256. • The sum of 64 sq. ft. and 256 sq. ft. is 320 sq. ft. • The square root of 320 sq. ft. carried to two decimal places is 17.88′ with 3056 remaining. Drop the remainder. • The length of the rafter is 17.88′.

The length of a common rafter per foot of run may be determined by applying the Pythagorean theorem. In a roof with a 1/4 pitch, the rafter rises 6″ for every 12″ of run. In the formula 6″ replaces a and 12″ replaces b. The square root of the sum of $a^2 + b^2$ gives the length of a common rafter per foot of run in inches.

Illustration. Find the length of a rafter per foot of run for a roof with a 1/4 pitch. The rise is 6″ and the run is 12″.

Solution. Use the formula $h = \sqrt{\text{rise}^2 + \text{base}^2}$.

Step 1. $h = \sqrt{6^2 + 12^2} = \sqrt{36 + 144}$

Step 2.

$$h = \sqrt{1'80.\,00'00\,00} = 13.42''$$

with quotient digits 1 3. 4 1 6

Step 3.

```
1 x 2 = 23 )        80
                    69
  2 x 13 = 264 )    11 00
                    10 56
 2 x 134 = 2681 )      44 00
                       26 81
2 x 1341 = 26826 )   17 19 00
                     16 09 56
                      1 09 44
```

Explanation. • To find the length of a rafter for one foot of run, multiply the rise (6″) by itself (6″ x 6″ equals 36 sq. in.). • Multiply the run (12″) by itself (12 x 12 equals 144 sq. in.). • Add the sums (36 sq. in. and 144 sq. in.) for a total of 180 sq. in. The square root of 180 sq. in. is 13.146 with 10944 remaining. • Round off to 13.42″.

APPLICATION

1. Determine the hypotenuse of a right triangle with a and b having the following measurements.

   a. a = 4′, b = 6′
   b. a = 6′, b = 6′
   c. a = 3′, b = 6′
   d. a = 2′, b = 8′
   e. a = 4′, b = 12′
   f. a = 4′, b = 8′
   g. a = 5′, b = 10′
   h. a = 6′, b = 11′
   i. a = 7′, b = 10′
   j. a = 7′, b = 9′

2. Find the length of the common rafter in figure 36-5. The run measures half the span.

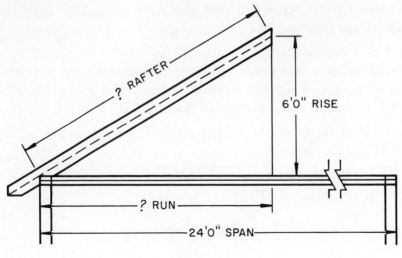

Fig. 36-5

3. Find the length of a common rafter on a roof with a rise of 8′ and a run of 18′.

4. What is the length of a common rafter on a roof with a rise of 7′ 4″ and a span of 32′ 6″?

5. The foundation layout in figure 36-6 is rectangular-shaped. Square the corner at point A.

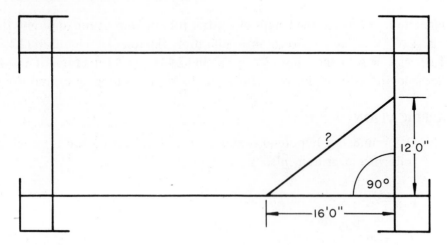

Fig. 36-6

6. A metal cap is 5″ square. What is the measurement of the diagonal of the square?

7. What is the length of a common rafter on a roof with a rise of 9′ and run of 14′?

8. The length of a stair stringer is determined in the same manner as the hypotenuse of a right triangle is found. The rise of the stairs in figure 36-7 is 8′ and the run is 9′. What is the length of the stringer?

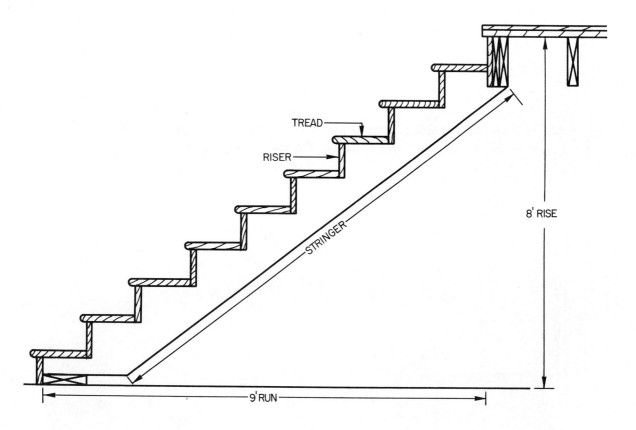

Fig. 36-7

CAREER PROFILE: THE MAINTENANCE CARPENTER

Job Description

Maintenance carpenters are involved in every aspect of carpentry work. These persons maintain existing structures such as colleges, hotels, and office buildings. Many are employed by government agencies. Since maintenance carpenters construct, install, and repair various parts of buildings, they must have a general knowledge of all aspects of carpentry, building practices, and safety regulations. The maintenance carpenter should be willing to handle various duties and emergency situations.

Qualifications

Preference is given to high school graduates or vocational trade school graduates. Various courses in carpentry are necessary for the maintenance carpenter to obtain general knowledge of the field. Good physical condition and stamina are necessary because of varied conditions.

Section 7
Fundamentals of Algebra

Unit 37
Introduction to Equations and Formulas

OBJECTIVES

After studying this unit, the student should be able to

- balance equations.
- write formulas.

Algebra is a branch of mathematics in which letters which represent certain numbers are combined to express different rules of arithmetic. These statements, known as *algebraic expressions,* may use two types of numbers, arithmetic numbers and literal numbers. *Literal* numbers are letters, such as X, a, or B, which represent specific numbers. *Arithmetic* numbers are the specific numbers that the letters represent, such as 2, 6, and 2/3.

The literal numbers may be used to express formulas, such as a + b = c. The literal numbers and arithmetic numbers may be used together to form equations. *Equations* are statements which show an equality or equivalence of expressions in mathematics, such as a + b = c + d.

It is important that the carpenter know the fundamentals of algebra so that he may make measurements and estimate materials more efficiently.

EQUATIONS

The information on ratio in Unit 20, Ratio, Proportion, and Percentage, may be helpful to the student when learning about equations. For example, the ratio which is expressed as 2:4::4:8 is an equation when it is expressed as 2/4 = 4/8.

Simple equations may be expressed by using the processes of addition, subtraction, multiplication, and division. Equations may also be used in determining square roots.

Addition

Illustration. Write the equation which shows that a certain factor may be added to 6 to equal 9, figure 37-1.

Solution. Let X represent the unknown number. The correct equation is X + 6 = 9.

Explanation. • The X is used to represent the unknown number which is added to 6 to equal 9. • The equation that balances the scale is X + 6 = 9.

Fig. 37-1 Balancing Equation

Subtraction

Illustration. Write the equation which shows that a number decreased by 5 will equal 6.

Solution. Let a represent the unknown number. The correct equation is $a - 5 = 6$.

Explanation. The a represents the unknown number. • The 5 is subtracted from the unknown number to equal 6. The equation, $a - 5 = 6$, is shown to be balanced when the symbol (a) is replaced with the number 11.

Multiplication

Illustration. Write an equation which shows that a certain number may be multiplied by 3 to equal 18.

Solution. Let x equal the unknown number. The correct equation is $x \cdot 3 = 18$.

Explanation. The x, which is the unknown in the equation, may be substituted to show a balanced equation.

Note: Since the symbol (x) is so frequently used in algebra as a literal number, multiplication of numerals is indicated by using a raised dot (·) between the number, such as $x \cdot 6$. Another method is to enclose one or both numbers in parenthesis, such as x (6) or (x)(6).

Division

Illustration. Write the equation which shows that a certain number may be divided by 4 to equal 5.

Solution. Let C equal the unknown. The correct formula is $C \div 4 = 5$.

Explanation. In this equation, C represents the unknown number. • When C is divided by 4, the result is 5. Therefore, $C \div 4 = 5$ is an equation because both sides of the equal sign are balanced. $C = 20$.

Square Root

Illustration. Write an equation which shows that a certain number may be added to the square root of 9 to equal 4.

Solution. Let X equal the unknown number. The correct formula is $\sqrt{9} + X = 4$.

Explanation. The square root of 9 plus an unknown (1) is 4. Therefore, the equation is balanced.

FORMULAS

Every letter in a formula represents something specific, while letters in equations represent an unknown. To solve a problem involving a formula, actual numerical values are substituted for each letter before the problem is solved.

A specific formula used in electricity, known as Ohm's Law, is stated as $I = E/R$. The letter I stands for current in amperes; E stands for voltage in volts; and R stands for resistance in ohms. The formula is an equation since I is equal to E/R, figure 37-2.

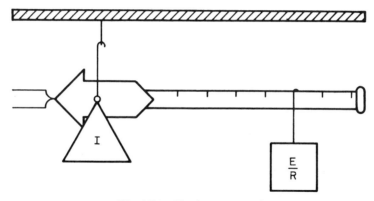

Fig. 37-2 Ohm's Law I = E/R

Illustration. Solve for E in the formula I = E/R if I represents .5 and R represents 6.

Solution.   I = E/R

   Step 1. I/1 = E/R
   Step 2. E = I x R
   Step 3. .5 = E/6
   Step 4. E = .5 · 6 = 3.0 voltage in volts

Explanation. • Consider the formula I = E/R as a proportion expressed as I/1 = E/R. • The problem then becomes E = I · R. • The letters are substituted for actual numbers which are multiplied.

APPLICATION

1. Write the following as equations and solve for the unknown.

   a. What number added to 6 equals 18?
   b. What number added to 5 equals 25?
   c. What number added to 4 equals 32?
   d. What number subtracted from 15 equals 7?
   e. What number subtracted from 12 equals 4?
   f. What number subtracted from 8 equals 2?
   g. What number multiplied by 5 equals 25?
   h. What number multiplied by 6 equals 18?
   i. What number multiplied by 4 equals 24?
   j. What number divided by 4 equals 4?
   k. What number divided by 5 equals 6?
   l. What number divided by 7 equals 9?

2. a. Write an equation which shows that a certain number may be added to the square root of 21 to equal 5.

   b. Write an equation which shows that a certain number may be added to the square root of 6 to equal 3.

   c. Write an equation which shows that a certain number may be added to the square root of 10 to equal 100.

3. The electrical circuit in figure 37-3 has four dry cell batteries. Each contains 1 1/2 volts and the resistance in ohms is 12. Write the formula and solve for the current in amperes.

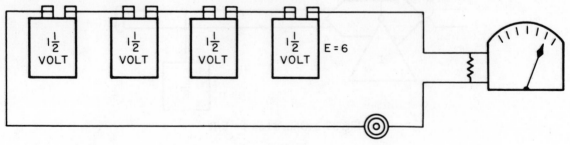

Fig. 37-3

4. Using the formula I = E/R, solve for E if I = .8 and R = 10.

5. Write a formula expressing each of the following.

a. area of a parallelogram
b. area of a circle
c. volume of a rectangular solid
d. area of a rectangle
e. volume of a cube
f. volume of a pyramid

Unit 38
Variables and Constants

OBJECTIVES

After studying this unit, the student should be able to

- identify variables and constants in formulas.
- identify direct variations in formulas.
- identify inverse variations in formulas.
- identify joint variations in formulas.

VARIABLES AND CONSTANTS

A number which is in a formula and whose value may change under different conditions is called a *variable.* The number in the formula whose value does not change is called a *constant.* For example, the formula for finding the volume of a right circular cylinder is $V = \pi r^2 h$. This formula can be used to find the volume of any right circular cylinder, whatever the value of r and h is. For each value of r and h, there is a value of V. When r and h change, V does also; but the value of π is always the same. The numbers V, r, and h are variables; the π is a constant. The value of V depends upon the value of r and h.

Illustration. Find the area of the circle in figure 38-1 whose radius is 2″.

Solution. $A = \pi r^2$
Step 1. $A = 3.1416 \times 2^2$
Step 2. $A = 3.1416 \times 4 = 12.5664$

Explanation. • The A and r are variables and the π is the constant in the formula. The value of 12.5664 depends upon the value of r.

The variations frequently occurring are direct variations, inverse variations, and joint variations.

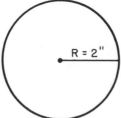

Fig. 38-1 Circle

DIRECT VARIATIONS

In some formulas, if either variable increases, the other increases; and if either decreases, the other decreases. In these cases, the two numbers are *direct variables* to each other.

The formula $I = E/R$, discussed in Unit 37, contains direct variables.

Illustration. Find the increase in current in amperes as the voltage in volts increases to 3 volts, figure 38-2A and B.

Solution.

I = E/R	I = E/R
I = 1.5/3	I = 3/3
I = .5	I = 1 (increase)

Explanation. This problem demonstrates direct variables since as the voltage increases (1.5 volts to 3), the current increases (.5 ampere to 1 ampere). Throughout the problem, resistance in ohms (3) remains constant.

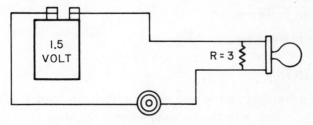

Fig. 38-2A Electrical Circuit

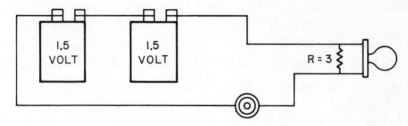

Fig. 38-2B Electrical Circuit Increase

INVERSE VARIABLES

In some formulas, as the value of one number increases, the value of the second number decreases. These are known as *inverse variables*. In an electrical circuit, the larger the wire is, the less the resistance is. The smaller the wire is, the greater the resistance is. The current flowing in the circuit depends upon the resistance.

Illustration. The resistance in a certain circuit is 6 and the voltage is 3. Show that the current decreases as the resistance increases.

Solution.

I = E/R
I = 3/6
I = .5

Explanation. As the resistance increases from 3 to 6, the voltage remains at 3 volts. Since the resistance is increased 2 times, the current is reduced by half, from 1 ampere to .5 ampere. Therefore, the current is an inverse variable to the resistance.

JOINT VARIABLES

In *joint variables,* one number varies jointly as the product of two or more numbers when it varies directly as the product of the other numbers.

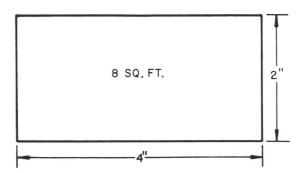

Fig. 38-3A Rectangular Solid

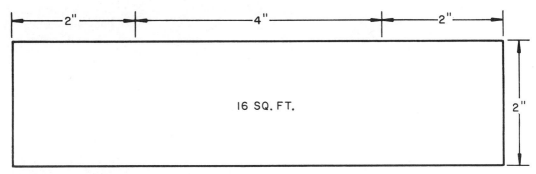

Fig. 38-3B Rectangular Solid

Illustration. The length of the rectangular solid in figure 38-3A is 4″ and the width is 2″. The area contains 8 sq. in. What is the area when the length is increased to 8″ and the width remains 2″, figure 38-3B? What is the area when the length is increased to 8″ and the width to 4″?

Solution.   Step 1. A = lw
                          A = 8″ x 2″ = 16 sq. in.
            Step 2. A = lw
                          A = 8″ x 4″ = 32 sq. in.

Explanation. In the first case, when the length is doubled to 8 and the width is constant at 2, the area is doubled and becomes 16 sq. in. When the length is 4″ and the width is doubled to 4″, the area is 16 sq. in. When the length and width are doubled to 8″ and 4″, the area is 32 sq. in.

APPLICATION

1. What is a variable?

2. Name three kinds of variables.

3. What is a constant?

4. Write a formula which contains a constant.

5. Write a formula which contains a variable and a constant.

6. A formula for finding board feet is t″ x w″ x L′/12. What is the constant?

7. The formula for finding the number of squares in an area is $W' \times L'/100$. What are the variables and the constant?

8. Indicate the constant to use for conversion in each of the following.

   a. square feet to square inches
   b. square inches to square feet
   c. square yards to square feet
   d. square feet to square yards

9. Indicate variables and constants in the following equations.

   a. $P = 2\,l + 2\,w$ e. $d = c/\pi$
   b. $P = 4S$ f. $c = \pi d$
   c. $d = 2r$ g. $c = 2\pi r$
   d. $r = d/2$ h. $c = \sqrt{a^2 + b^2}$

Unit 39
How to Transpose Equations

OBJECTIVES

After studying this unit, the student should be able to

- transpose equations by moving numbers from the left side of the equation to the right side and from the right to the left side.

- solve equations by addition, subtraction, multiplication, and division.

To find the unknown in an equation or formula, transposing may be used. When a term is *transposed,* it is moved from one side of an equation or formula to the other side. This process always involves a change of operation. Other ways of transposing involve applying an operation, such as addition, subtraction, multiplication, or division to both sides of the equation or formula so that a term is cancelled.

The unknown must remain alone on one side of the equation or formula when transposing a term from one side to the other. It may be necessary to move all other quantities to the opposite side. When a quantity is transposed from one side of the equation to the other side, change the sign from plus to minus or from minus to plus.

Illustration. The total length of the block in figure 39-1 is 14″. One section measures 2″. What is the length of the other section?

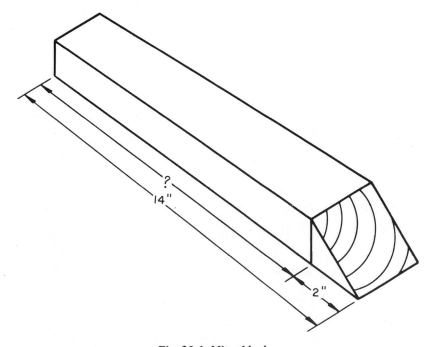

Fig. 39-1 Miter block

Solution.   Let x equal the unknown length in inches.

Step 1.	x + 2 = 14″
Step 2.	x = 14 – 2
Step 3.	x = 12″

Check. x + 2 = 14

x = 14 – 2

x = 12

12 + 2 = 14

Explanation. • Move the 2 to the right side of the equation while the x remains on the left. • The +2 placed on the right side is changed to a –2. The equation is x = 14 – 2. • Subtracting 2 from 14 leaves 12. The x = 12″. To prove the answer, substitute 12 for x in the equation and solve.

Sometimes the unknown number is on the right side of the equation.

Illustration. The length of the block in figure 39-1 is expressed in the equation 14″ = x + 2″.

Solution.   Step 1.   14″ = x + 2″   Check. 14″ = x + 2″

Step 2.   14″ – 2 = x   14″ = 12″ + 2″

12 = x

Explanation. • The 2 is transposed from the right side of the equation to the left side. • The x remains on the right side. The answer is 12″.

OTHER METHODS OF SOLVING EQUATIONS

Addition

Another method of solving equations involves adding the same number to both sides of the equation.

Illustration. *Joint cement*, or *spackle*, is combined with water to produce a mixture which is used to treat joints in gypsum-wallboard construction. *Oakum* is hemp or rope which is used to caulk joints.

A carpenter has a jar of joint cement which weighs 3 lb. What is the weight of oakum if the joint compound weighs 2 lb. less than the oakum?

Solution. Let x equal the weight of the oakum.

x – 2 = weight of joint cement in pounds.

Step 1.   x – 2 lb. = 3 lb.

Step 2.   2 – 2 lb. + x = 3 lb. + 2

Step 3.   x = 3 + 2

x = 5 lb.

Check.   5 – 2 = 3

Explanation. • Add 2 to x – 2 on the left side of the equation. Subtract 2 from 2, which equals x. • Adding 2 to 3 on the right side of the equation equals 5. This shows that the weight of the lead is 5 lb. • To prove that x equals 5, substitute 5 for x and solve.

Subtraction

If the same quantity is subtracted from both sides of the equation, the equation remains balanced.

The process of subtraction is used when the equation contains the sum of two or more quantities.

Illustration. The total length of the pipe layout in figure 39-2 is 16′. The sections measure 2′, 3′, and x′. What is the length of x?

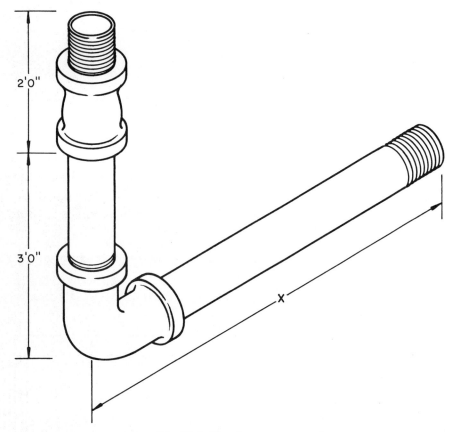

Fig. 39-2 Pipe Layout

Solution. Let x equal the unknown length.

Step 1. $2' + 3' + x' = 16'$ (equation)

Step 2. $5 + x = 16'$

Step 3. $5' - 5' + x' = 16' - 5'$
$x = 11'$

Check. $2' + 3' + 11' = 16'$ $\qquad$ $5' + 11' = 16'$

Explanation. • The sum of the known numbers on the left side of the equal sign is 5. • Subtract 5 from both sides of the equal sign. In this example, x equals 11′. • To be certain that 11′ is the correct length, substitute 11′ in the equation. • The result is 16′ on each side of the equal sign.

Multiplication

Both sides of the equation may be multiplied by the same number without changing the balance. The process of multiplication is used when the equation contains a fraction. All terms on the right side of the equation and all terms on the left side of the equation must be multiplied by the same multiplier.

Illustration 1. The measurement of the pipe in figure 39-3A is expressed by the equation $a/4 = 9''$. Find the value of a.

Solution. $a/4 = 9$ (equation) Check. $a/4 = 9$

$$\frac{\overset{1}{\cancel{4}} \cdot a}{\underset{1}{\cancel{4}}} = 9 \cdot 4$$

$$a = 36$$

$a = 36$

$36/4 = 9$

Explanation. The unknown that is represented by a is found by multiplying both sides of the equation by 4. On the left side of the equal sign, each 4 is cancelled, leaving $1/1a$, or a. On the right side of the equal sign, 9 multiplied by 4 equals 36; a equals 36. Substitute 36 in the place of a and divide 36 by 4, which equals 9. Both sides of the equation are equal.

Illustration 2. The measurement of the pipe in figure 39-3B is expressed by the equation $a/4 + 2 = 9$. Find the value of a.

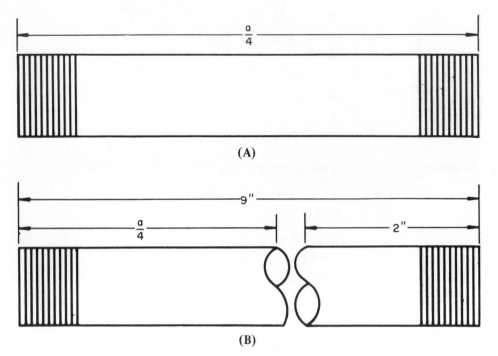

(A)

(B)

Fig. 39-3 Water Pipes

Solution.   a/4 + 2 = 9   Check.   28/4 + 2 = 9

Step 1. $\dfrac{4 \cdot a}{4}$ + 4 · 2 = 4 · 9

7 + 2 = 9

9 = 9

Step 2. a/1 + 8 = 36

Step 3.   a + 8 = 36
Step 4.   a = 36 − 8
          a = 28

Explanation. • Multiply each term or quantity on both sides of the equal sign by 4. • The 4 multiplied by a/4 equals a/1 and 4 multiplied by 2 equals 8. All of these quantities are placed on the left side of the equal sign. • On the right side of the equal sign, multiply 9 by 4, which equals 36. • Change the 8 to −8, transpose it to the right side of the equal sign, and subtract, leaving 28. In this problem, a is equal to 28. • Substitute 28 in place of a and a/4 = 28/4. Then 28/4 equals 7. • Add 7 and 2 which equals 9. The final equation balances, with 9 on each side of the equation.

Division

Both sides of the equation may be divided by the same quantity without changing the balance.

Illustration. The length of the box in figure 39-4 is 30″. The box is divided into 5 equal sections. How long is each section?

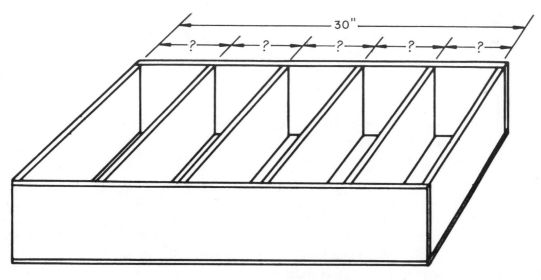

Fig. 39-4 Compartment Box

Solution. 5x″ = 30″   Check.   5x = 30
                                                    5 · 6 = 30
                                                    30 = 30

Step 1.

$$\frac{\overset{1}{\cancel{5}x}}{\underset{1}{\cancel{5}}} = \frac{\overset{6}{\cancel{30}}}{\underset{1}{\cancel{5}}}$$

Step 2. $x/1 = 6/1$

$x = 6''$

Explanation. • Divide numbers on both sides of the equal sign by 5; x is 6. • Substitute 6 in the place of x and multiply 5 by 6. • The number on both sides of the equation is 30.

APPLICATION

1. Solve each of the following by subtracting the same number from both sides of the equation. Check each answer.

   a. $x + 5 = 8$ e. $a + 8 = 10$ i. $8 = x + 1$
   b. $x + 2 = 8$ f. $b + 6 = 10$ j. $8 = a + 2$
   c. $x + 2 = 9$ g. $b + 5 = 12$
   d. $a + 7 = 9$ h. $t + 5 = 11$

2. Solve each of the following equations by multiplying and for some equations, subtracting equation by the same numbers. Check each answer.

   a. $x/2 = 8$ e. $a/9 = 5$ i. $x/9 + 4 = 6$
   b. $x/3 = 9$ f. $x/4 + 2 = 8$ j. $x/3 + 3 = 6$
   c. $a/4 = 4$ g. $a/6 + 4 = 6$
   d. $a/4 = 12$ h. $a/5 + 2 = 10$

3. Solve each of the following equations by transposing a term from one side of the equation to the other side. Check each answer.

   a. $x + 2 = 8$ e. $s + 8 = 14$ i. $b + 10 = 16$
   b. $a + 7 = 8$ f. $8 + a = 6$ j. $c + 5 = 5$
   c. $b + 2 = 9$ g. $g + 4 = 9$
   d. $s + 4 = 10$ h. $g + 7 = 10$

4. Solve each of the equations by adding the same number to both sides of the equation. Check each answer.

   a. $a - 3 = 10$ e. $s - 5 = 3$ i. $g - 12 = 6$
   b. $x - 5 = 15$ f. $s - 6 = 9$ j. $g - 7 = 10$
   c. $x - 4 = 12$ g. $h - 7 = 20$
   d. $h - 7 = 2$ h. $g - 8 = 12$

5. Solve each of the following equations by dividing both sides of the equation by the same number. Check each answer.

   a. $4x = 12$ e. $5a = 15$ i. $24 = 4x$
   b. $9x = 18$ f. $3a = 21$ j. $4h = 20$
   c. $6a = 24$ g. $6x = 48$
   d. $7a = 14$ h. $10 = 5c$

Unit 40
How to Determine Letter Value in Formulas

OBJECTIVE

After studying this unit, the student should be able to

- solve for letter value in formulas.

SOLVING FOR LETTER VALUES IN SIMPLE FORMULAS

Both large and small letters are used in formulas. The large letters and small letters have different meanings, such as the letters A and a in the formula A = ab. Each letter represents quantities. They are connected by mathematical signs, such as (+), (−), (÷) or the multiplication sign (·).

Illustration. The formula for finding the area of a parallelogram is A = ab. Find the value of A with the dimensions given in figure 40-1.

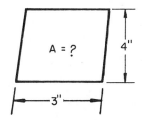

Fig. 40-1 Parallelogram

Solution. A = ab   A = 4″ · 3″   A = 12 sq. in.

Explanation. • The first step in solving the formula is to substitute 4 in the place of a and 3 in the place of b. • The ab in the formula indicates multiplication. A = 4″ · 3″ which equals 12 sq. in.

The value of the other letters in the formula may be found in the same way.

Illustration. Find the value of a in the formula A = ab, figure 40-2.

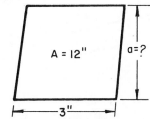

Fig. 40-2 Parallelogram

Solution. A = ab

Step 1.   12 = a · 3

Step 2. $\dfrac{12}{\cancel{3}_{1}} = a \cdot \dfrac{\cancel{3}^{1}}{\cancel{3}_{1}}$

$4'' = a$

Explanation. • Substitute numbers for letters and solve. $a = 4''$.

In the same way, by knowing the value of the letters A and a, the value of b may be found in the formula A = ab.

Illustration. In the formula A = ab, find the value of b, figure 40-3.

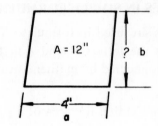

$A = 12''$? b

4"
a

Fig. 40-3 Parallelogram

Solution. A = ab

Step 1. 12 = 4b

Step 2. b = A/a

Step 3. $\dfrac{\cancel{12}^{3}}{\cancel{4}_{1}} = \dfrac{\cancel{4}b}{\cancel{4}_{1}}$

$3'' = b$

Explanation. • Following the same procedure, divide each side by 4. The answer is 3''.

The hammer, a tool used extensively in carpentry, must be handled properly for best results, figure 40-4. The force with which a hammer strikes an object can be determined by the formula force = weight x applied force x length of handle.

Note: The weight of a hammer is determined by the weight of its head.

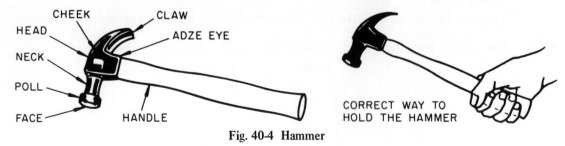

CHEEK   CLAW
HEAD   ADZE EYE
NECK
POLL
FACE   HANDLE

CORRECT WAY TO
HOLD THE HAMMER

Fig. 40-4 Hammer

Illustration. The weight of a certain hammer is 14 oz. The handle length is 12''. If a carpenter applies 11 lb. of force, with how many pounds of force will the hammer strike the object?

Solution. force = wt. • a • l

Step 1.   14 oz. = 7/8 lb.

Step 2. force = 7/8 lb. x 11 lb. x 12

Step 3. force = 77/8 lb. x 12
force = 115.5 lb.

Explanation. • The weight of the hammer is 14 oz. • Change 14 oz. to 7/8 lb. • Multiply 7/8 lb. by 11 lb., which equals 77/8. • Multiply 77/8 by 12, which equals 924/8. • Divide 924 by 8 to determine the striking force of the hammer, 115.5 lb.

FORMULAS CONTAINING PARENTHESES AND BRACKETS

When parentheses are used in formulas, the expressions contained in the parentheses are solved first. The brackets are then removed and the entire problem is solved.

Illustration. Solve the following formula.

$$R = \frac{[(a-b) + (a-b)^2 - (b-1) + (b-1)^2] \; D}{2a} + \frac{s + (b-1) + (b-1)^2}{2a}$$

Solution. Let a = 8, b = 2, s = 500, D = 1000, R = unknown

Step 1. Substitute actual values in the formula in place of letters.

$$R = \frac{[(8-2) + (8-2)^2 - (2-1) + (2-1)^2] \; 1000}{2 \times 8} + \frac{500 + (2-1) + (2-1)^2}{2 \times 8}$$

Step 2. Perform the procedures inside of the parentheses.

$$R = \frac{[(6) + (6)^2 - (1) + (1)^2] \; 1000}{16} + \frac{500 + 1 + (1)^2}{16}$$

Step 3. Remove parentheses on the inside of the brackets.

$$R = \frac{[6 + 36 - 1 + 1] \; 1000}{16} + \frac{500 + 1 + 1}{16}$$

Step 4. Solve the problem inside the brackets.

$$R = \frac{[42] \; 1000}{16} + \frac{502}{16}$$

Step 5. Remove the brackets and add.

$$R = \frac{42000}{16} + \frac{502}{16} = \frac{42502}{16} = 2656.37$$

Explanation. • In the illustration, problems within parentheses are solved before problems within brackets are solved. • By Step 5, the problem has become simple addition. • The answer is converted to a decimal.

APPLICATION

1. Consider the formula bd. ft. $= \dfrac{t'' \text{ x } w'' \text{ x } 1'}{12}$. Find the number of bd. ft. if t = 1", w = 6", and 1 = 12'.

2. A carpenter must determine the width of a room whose area is 368 sq. ft. and the length 23′. The formula is A = lw.

3. A carpenter has a board which has an area of 126 sq. in. The width is 9″. What is its length? The formula is A = lw.

4. The total area of a cylinder with a radius of 21″ is 6,732 sq. in. What is the height of the cylinder in inches? The formula is $A = 2\pi rh + 2\pi r^2$.

5. In the formula P = 2 1 + 2 w, solve for P if l equals 6′ and w equals 4′.

6. The circumference of a circle is equal to pi (π) times the diameter. The circumference measures 18.8496″. What is the diameter?

7. In the formula C = 5/9 (F-32), find F if C equals 0.

8. Solve the formula $V = 4/3\pi r^3$ if r = 3.

9. Find the length of a hip rafter (LH) whose run is 6′ and length per foot run (1) is 18″. The formula is LH = R x 1/12.

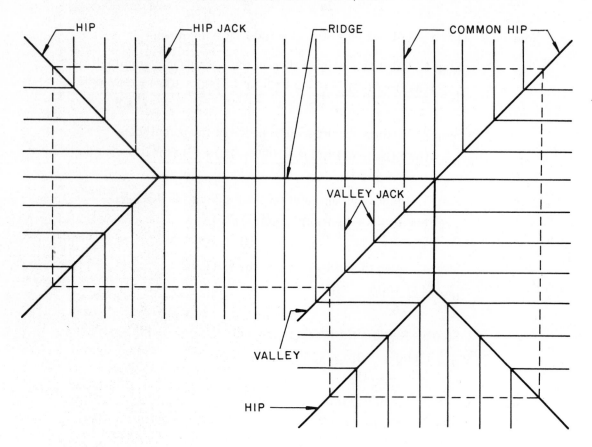

10. In the formula P = 2 1 + 2 w, find P if l equals 16″ and w equals 14″.

11. In the formula V = lwh, find l if V equals 64 cu. in., w equals 4″, and h equals 8″.

12. In the formula F = 9/5 C + 32, find F if C equals 100.

13. In the formula G = 7 1/2 lwh, find G if l equals 15', w equals 10', and h equals 5'.

14. In the formula A = 1/(1/a + 1/b), find A if a equals 2 and b equals 3.

15. In the formula R = 1/(1/a + 1/b + 1/c), find R if a equals 4, b equals 6, and c equals 8.

16. In the formula I = nE/(Re +Ri), find I if n equals 20, E equals 1.1, Re equals 200, and Ri equals 4.

17. The floor area of a certain room is 144 sq. ft. and the width is 12'. What is the length?

18. The floor area of a certain room is 400 sq. ft. and the width is 20'. What is the length?

19. Use the formula A = lw to find the unknowns as indicated.

    a. A = 15 sq. ft., w = 10', and l = ?
    b. A = 121 sq. ft., l = 11', and w = ?
    c. A = ?, l = 13 ft., and w = 11'

20. Use the following formula to find the value of P if a = 6, x = 4, Y = 1500, and r = 1000.

$$P = \frac{[(a\text{-}x) + (a\text{-}x)^2 - (x\text{-}2) + (x\text{-}2)^2]\ Y}{3a} + \frac{r\ (x\text{-}2) + (x\text{-}2)^2}{3a}$$

Unit 41
How to Solve Equations by Square Root

OBJECTIVES

After studying this unit, the student should be able to

- find the square root of an algebraic expression.
- find the square root of a fraction.

Every number can be considered to have two square roots. For example, +4 and −4 are both square roots of 16, since $(+4) \cdot (+4) = +16$ and $(−4) \cdot (−4) = +16$.

The plus sign (+) (also called the *positive sign*) before a number shows that it is above zero on the number scale. When a reading is below zero, the *negative sign* (−) is used.

The radical written with a number but without a sign indicates that the square root of the number to be taken is positive, such as $\sqrt{4}$ or $\sqrt{x^2}$. When the negative square root of a number is to be determined, the minus sign is placed before the radical, such as $-\sqrt{4}$ or $-\sqrt{x^2}$.

FINDING THE SQUARE ROOT OF AN ALGEBRAIC EXPRESSION

The square root of an algebraic expression can be found in the same manner as the square root of an arithmetic number.

Illustration. The square figure 41-1 contains 2.25 sq. in. Find the length of one side.

Solution. $A = S^2$

$S^2 = A$

Step 1. $S = \sqrt{A}$

Step 2.
$$S = \sqrt{2.'25}$$
$$\begin{array}{r} 1.5 \\ \hline 1 \\ 25)\,1\ 25 \\ \underline{1\ 25} \end{array}$$

$S = 1.5$ in.

Proof. $S^2 = 1.5$ in. x 1.5 in.

$S^2 = 2.25$ sq. in.

225 SQ. IN.

Fig. 41-1 Square

Explanation. The area of the square is 2.25 sq. in. • Find the length of its side by finding the square root of the area. The formula is $A = S^2$. • If $A = S^2$, then $S^2 = A$ and S equals $\sqrt{A}$. • Place 2.25 under the square root sign. The square root of 2.25 sq. in. is 1.5 in., which is the length of each side. • To prove the answer, multiply 1.5 in. x 1.5 in., which equals 2.25 sq. in. The answer is 1.5 in.

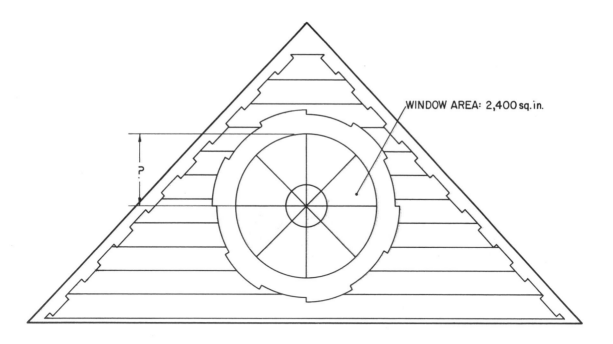

WINDOW AREA: 2,400 sq. in.

Illustration. Find the radius of a circle whose area measures 2400 sq. in.

Solution. A = πr^2

A = 2400 sq. in.

π = 22/7

Step 1. r^2 = 2400 ÷ 22/7

r^2 = 2400 x 7/22 = 16800/22

Step 2. r^2 = 763.6 = 764

Step 3.

$$r = \sqrt{7'\,64'.\,00\ 00}$$

```
                 2  7 . 6  4
        r =   √ 7' 64'. 00 00
                 4
        47 ) 3 64
             3 29
      546 )    35 00
              32 76
     5524 )    2 24 00
               2 20 96
                  3 04
```

Answer r = 27.64+"

Explanation. Determine the radius by finding the square root of the result of 2400 ÷ 22/7 (763.6 or 764 rounded off). Find the square root of 764 by the usual method. The radius is 27.64+".

Note. The symbol (+) written after 64 in the answer indicates a remainder.

The length of the side of an equilateral triangle is given by the formula $h = (S/2)\sqrt{3}$. Therefore, the length of a rafter which forms an equilateral triangle, figure 41-2, can be determined by the same formula as long as the three sides are equal.

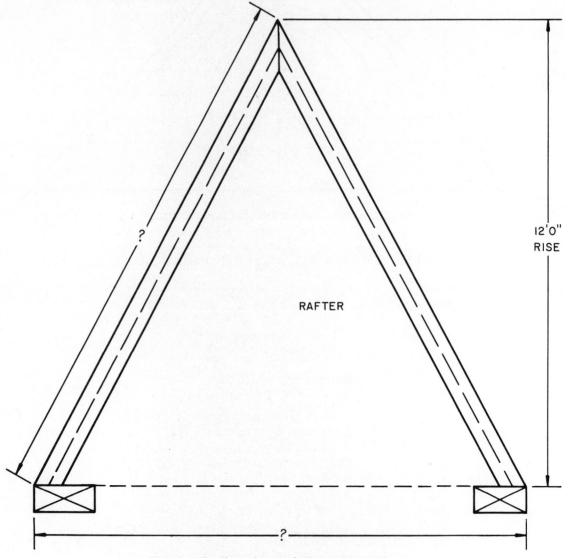

Fig. 41-2 Equilateral Triangle Formed by Rafters

Illustration. If S is the length of a rafter in the figure, the rise (h) is determined by the formula $h = (S/2)\sqrt{3}$. Find S if the rise (h) is 12′.

Solution. $h = (S/2)\sqrt{3}$

$12' = (S/2)\sqrt{3}$

Step 1. $S\sqrt{3} = 2h$

Step 2. $S\sqrt{3} = 2 \times 12' = 24$

Step 3. $S = 24/\sqrt{3}$

Step 4.
$$\begin{array}{r} 1.\ 7\ 3\ 2 \\ \sqrt{3.'\,00'\,00\,00} \\ 1 \\ \hline 27\ )\ 2\ \ 00 \\ 1\ \ 89 \\ \hline 343\ )\quad 11\ \ 00 \\ 10\ \ 29 \\ \hline 3462\ )\qquad 71\ 00 \\ 69\ 24 \\ \hline 1\ 76 \end{array}$$

Step 5.
$$\begin{array}{r} 13.856 \\ 1.732.\overline{)\,24000.000} \\ 1732 \\ \hline 6680 \\ 5196 \\ \hline 1484\ 0 \\ 1385\ 6 \\ \hline 98\ 40 \\ 86\ 60 \\ \hline 11\ 800 \\ 10\ 392 \\ \hline 1\ 408 \end{array}$$

S = 13.856 or 13.86

Explanation. • Substitute 12′ in the place of h in the formula. • The square root of 3 is 1.732. • Divide 24′ by 1.732, which equals 13.856. Rounded off, 13.856 becomes 13.86. The answer is 13.86′.

FINDING THE SQUARE ROOT OF A FRACTION

The square root of a fraction may be determined by dividing the square root of the numerator by the square root of the denominator.

Illustration. Find the square root of the fraction 1/4.

Solution. $\sqrt{1/4} = \dfrac{\sqrt{1}}{\sqrt{4}} = 1/2$

Explanation. The square root of the numerator (1) is 1. The square root of the denominator (4) is 2. • Divide 2 by 1 which equals .5. • To check the answer, multiply .5 by .5 which equals .25; .25 = 1/4.

APPLICATION

1. Find the square root of the following fractions to the nearest hundredth. Use the square root table in the Appendix when possible.

   a. 3/4   j. 7/8
   b. 2/3   k. 7/12
   c. 4/12   l. 4/15
   d. 4/16   m. 4/5
   e. 4/15   n. 2/5
   f. 9/25   o. 5/9
   g. 5/8   p. 4/9
   h. 9/16   q. 15/32
   i. 5/16   r. 25/64

2. Find the radius of the circle in figure 41-3 whose area is 1500 sq. in.

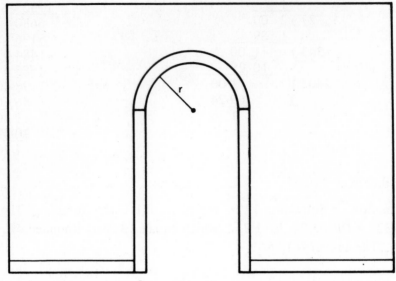

Fig. 41-3

3. One formula for finding the area of a circle is A = (1/4) πd^2. Find the diameter of the circular louver in figure 41-4 whose area measures 144 sq. in. In solving the problem, π equals 22/7. *Note:* A *louver* is an opening used to ventilate closed areas, such as attics.

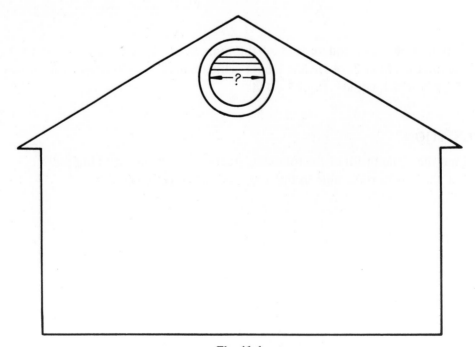

Fig. 41-4

4. The area of a certain cube is 54 cm^2. Find the length of its sides in centimeters.

5. The surface area of a certain sphere is 314.15 sq. ft. What is the radius of the sphere?

6. The formula V = (1/3) πr^2 h is used to find the volume of a right circular cone. A certain container has volume of 75.4 cu. in. and a height of 6''. What is the radius of the container?

CAREER PROFILE: THE CONTRACTOR

Job Description

Contractors are the persons who assume the entire responsibility of carrying out the plans of designers and architects. They estimate, bid, and sign agreements to construct or repair buildings. Contractors are responsible for following the general conditions of a contract. Workers on a job are employed and paid by the contractor.

There are two types of contractors in carpentry, general and special. General contractors usually contract to construct a complete building. Special contractors usually subcontract to work in a specialized field of carpentry, such as cabinetmaking or form building. Contractors have the advantage of being independent business persons. In this respect, contractors exercise a great amount of control over their working conditions. Their work is done both indoors and outdoors and in all geographical locations.

Qualifications

Contractors must possess a general knowledge of all facets of construction, such as characteristics of materials, safety practices, trade theory, and interpretation of blueprints and specifications. Many receive special training in bidding, securing bonds, and reading contracts. A contractor's license is required. A construction worker may spend years working as a journeyman before applying for a license.

In the classroom and on the job, the contractor must learn the relationship between carpentry and the other building trades.

Example of Signing of the Contract to Initiate the Construction of a Building

Section 8
Fundamentals of Geometry

Unit 42
Circles and Polygons

OBJECTIVES

After studying this unit, the student should be able to

- define regular, concentric, and eccentric circles.

- construct certain regular polygons.

- determine the radius of a circle.

Geometry is a branch of mathematics dealing with the measurement, relationship and properties of figures, points, and solids. One of the most commonly studied figures in geometry is the circle.

A circle contains 360°, with all points an equal distance from the center, figure 42-1. Many objects in carpentry work are based on the circle, such as bolts, nuts, pipes, and some architectural construction.

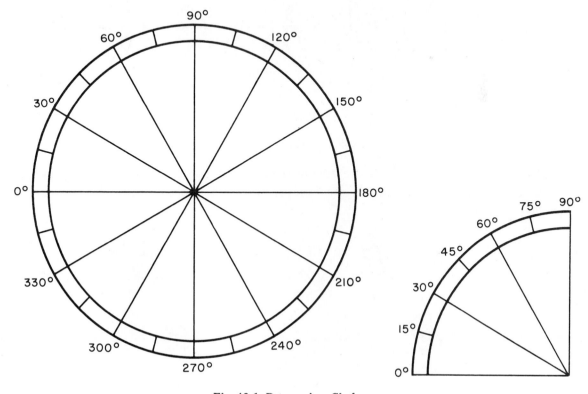

Fig. 42-1 Degrees in a Circle

243

There are three basic types of circles: regular, concentric, and eccentric, figure 42-2. *Concentric* circles have a common center, while eccentric circles have different centers.

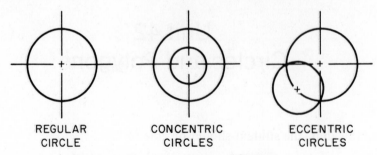

REGULAR CONCENTRIC ECCENTRIC
CIRCLE   CIRCLES   CIRCLES

Fig. 42-2 Three Basic Types of Circles.

This unit requires the use of such instruments as a divider, compass, rule or scale, and straightedge.

REGULAR POLYGONS

A *polygon* is a geometrical figure or object which is bound by straight lines. In a *regular polygon,* the lines are equal and the angles are equal, figure 42-3.

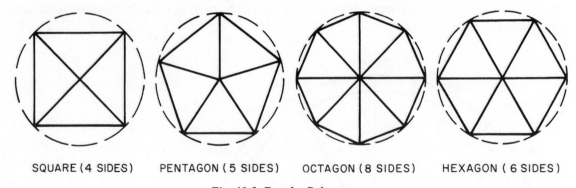

SQUARE (4 SIDES) PENTAGON (5 SIDES) OCTAGON (8 SIDES) HEXAGON (6 SIDES)

Fig. 42-3 Regular Polygons

Bolts and nuts, figure 42-4, are used in carpentry for fastening purposes. The nuts and heads of the bolts may be considered polygons, since they may be round, hexagonal, or square in shape.

MACHINE   FLAT HEAD
BOLT   STOVE BOLT

Fig. 42-4 Commonly Used Nuts and Bolts

Illustration. Draw a layout for a bolt with a square head and a 1″ diameter.

Solution.   Step 1.

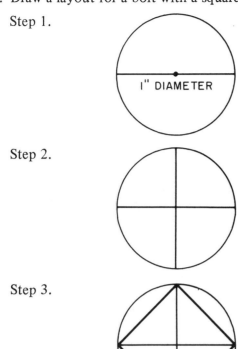

Step 2.

Step 3.

Explanation. • Draw a circle 1″ in diameter with a compass. • Draw two lines through the center and at right angles to each other using a 45° triangle. Extend the lines to intersect the circumference. Connect the four points where the lines intersect the circumference.

It may be necessary for the carpenter to locate the center of a round object or figure. The combination square in figure 42-5 has a center square which may be used for this purpose. The center square is designed so that any two lines drawn across the end of a round object will cross its center.

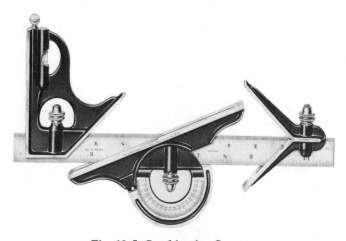

Fig. 42-5 Combination Square

The center of a circle may also be determined by bisecting lines connecting points on the circle. *Bisecting* involves dividing an object or figure into equal parts.

Illustration. Find the center of the circle with an unknown diameter.

Solution.

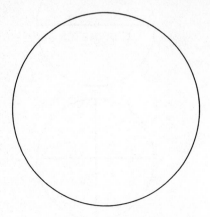

circle with unknown diameter

Step 1. Mark three arbitrary points (A, B, and C) on the circumference of the circle.

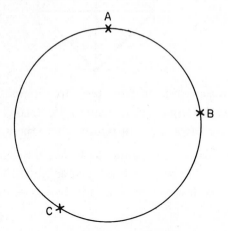

Step 2. Connect points A and B and B and C with straight lines.

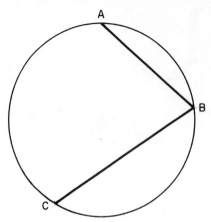

Step 3. From points A, B, and C, draw equal arcs with a radius greater than half AB and BC. Extend perpendicular bisectors DE and FG to intersect at center point O.

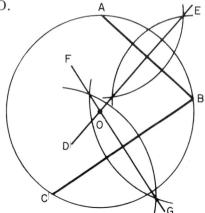

Explanation. • Mark the three arbitrary points (A, B, and C) on the circumference with a compass. Connect points A and B and B and C with straight lines. • Set the compass so that the radius is larger than the radius of the circle. • Bisect lines AB and BC. • Extend the bisectors DE and FG. The intersection at point O is the center of the circle.

FINDING THE RADIUS OF A CIRCLE

The radius, equal to one-half the diameter of a circle, extends directly from the center point of the circle to its circumference. If the area of a given circle is known, its radius can be determined by the formula $r = \sqrt{\dfrac{A}{\pi}}$.

Illustration. The area of a certain circle is 12 sq. ft. Find the radius of the circle. Round off pi in determining the answer.

Solution. $r = \sqrt{\dfrac{A}{\pi}}$

Step 1. $\dfrac{12}{3.1416}$

Step 2.
```
                3.8197
3.1416.) 120000.0000
         94248
         25752 0
         25132 8
           619 20
           314 16
           305 040
           282 744
            22 2960
            21 9912
               3048
```

Step 3.
```
           1.  9   5   4   4
(4)√3.' 81' 97' 00' 00
    1
2 x 1 = 29 ) 2  81
             2  61
2 x 19 = 385 )  20  97
                19  25
2 x 195 = 3904)  1 72  00
                 1 56  16
2 x 1954 = 39084)   15 84  00
                    15 63  36
                       20  64
```

Answer = 1.9544′

Explanation. • In the formula, substitute 12 sq. ft. for A. • Divide 12 by 3.1416, which equals 3.8197. The square root of 3.8197 is 1.9544′ which is the radius of the circle.

The Octagon

An *octagon* is a geometrical figure or object with eight sides. A *regular octagon* has eight equal sides and eight equal angles. It may be constructed from a given circle.

Illustration. Construct a regular octagon from a given circle.

Solution.

Step 1.   Step 2.

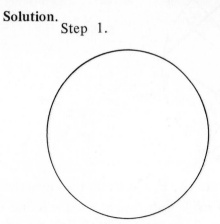 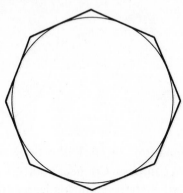

Explanation. • Draw eight equal sides with equal angles tangent to the circle with a T-square and 45° triangle. A line is *tangent* to the circle when it touches the circumference without going through it.

The Pentagon

The *pentagon* is a geometrical figure or object with five equal sides. A *regular pentagon* has five equal sides and five equal angles. The figure is often used in construction. It is constructed by the same basic principle as a hexagon.

Illustration. Draw a circle and circumscribe a regular pentagon in the circle.

Solution and

Explanation.   Step 1. Draw a circle with a diameter equal to the approximate size of the pentagon's sides. Let AB be the horizontal centerline and CD the vertical centerline. Point O marks the center of the circle.

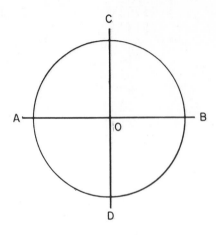

Step 2. With the point of the compass first on O and then on B, bisect OB with two arcs. Locate point E midway between points O and B. The setting on the compass should be greater than half the radius of the circle but not greater than the radius.

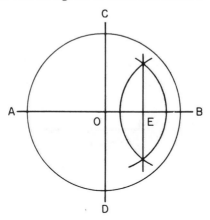

Step 3. Set the compass to measure the distance of EC and draw an arc intersecting AO at point F.

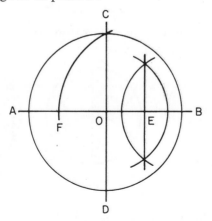

Step 4. Set the compass to measure the distance of FC and, following around the circle, locate five points on the circumference.

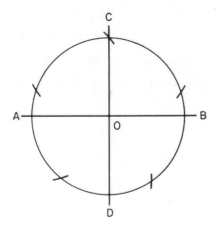

Step 5. With a straightedge, connect the points with straight lines to complete the pentagon.

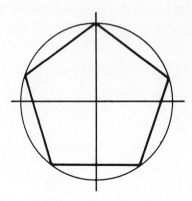

APPLICATION

1. Define eccentric and concentric circles.

2. With drawing instruments, draw an equilateral triangle inside a circle.

3. With drawing instruments, draw a square inside a circle.

4. With drawing instruments, draw a hexagon using arcs equal to the radius to locate all points, figure 42-6.

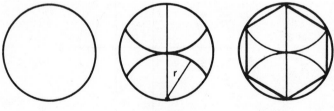

Fig. 42-6

5. Inscribe a regular pentagon in a circle. To *inscribe* is draw a figure within another figure so that the edges of the two touch in as many places as possible.

6. Inscribe a regular octagon in a circle.

7. Draw a 1″ square. Locate its center and inscribe a circle.

8. A round bolt head has an area of 9.4248 sq. in. Find its radius.

9. The cross-sectional area of a pipe is 31.4160 sq. in. What is the radius of the pipe?

Unit 43
Angles

OBJECTIVES

After studying this unit, the student should be able to

- convert measurements of angles of one unit to another unit.
- lay out and measure angles.
- read a bevel protractor.
- add, subtract, multiply, and divide angle measurements.

Angles are an important part of many jobs in carpentry, such as laying out and reading building plans. It is therefore necessary for the carpenter to have a thorough knowledge of angles and their construction.

Angles, sometimes expressed by the symbol ∠, are measured in degrees and fractional parts of a degree. The degree is designated by the symbol (°) placed to the right and a little above the number. One degree represents 1/360 of a complete circle. The largest fractional part of a degree is the *minute*, designated by the symbol ('). It represents 1/60 of a degree. The smallest fractional part of a degree, the *second*, is designated by the symbol ('') and represents 1/60 part of a minute.

Figure 43-1 gives degrees and equivalents.

Figure Degrees and Equivalents

360 degrees (°) = 1 circle

60 minutes (') = 1 degree

60 seconds ('') = 1 minute

Fig. 43-1

An angle is formed when two lines meet at one point. The point at which they meet is called the *vertex*.

Angles are measured by the opening between the two lines, figure 43-2. The length of the sides has nothing to do with the measurement of the angle.

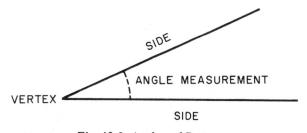

Fig. 43-2 Angle and Parts

There are special tools designed to lay out angles, such as the framing square, try square, and combination square. The protractor is also commonly used.

CONVERSION OF ANGLE MEASUREMENT

Changing Degrees to Minutes and Minutes to Degrees

Changing degrees to minutes requires the process of multiplication.

Illustration. The threads on the bolt in figure 43-3 form 60° angles to each other. Change 60° to minutes.

Solution. 1 degree = 60 minutes

$$\begin{array}{r} 60 \\ \times\ 60 \\ \hline 3600\ \text{minutes} \end{array}$$

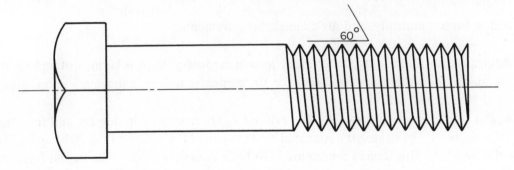

Fig. 43-3 Bolt Threads at 60° Angle

Explanation. Since a degree equals 60 minutes, multiply 60 by 60 minutes/degree. The answer is 3600 minutes.

Changing minutes to degrees requires division.

Illustration. An *angle iron* is a piece of iron in the shape of a right angle. The angle iron in figure 43-4 measures 2700 minutes. Change this measurement to degrees.

Solution. 60 minutes equals 1°

$$2700' \div 60\ \frac{\text{min.}}{\text{degree}} = 45°$$

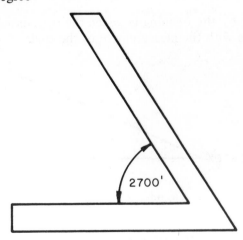

Fig. 43-4 Angle Iron

Explanation. There are 60 minutes in 1° so divide 2700′ by 60. The answer is 45°.

Changing Minutes to Seconds and Seconds to Minutes

Illustration. A protractor reads 20° 50′. Change 50′ to seconds. *Note:* This measurement indicates an *acute angle,* one greater than 0° and less than 90°. An *obtuse angle* is one which measures greater than 90° and less than 180°. A *reflex angle* measures greater than 180° and less than 360°.

Solution. 1 minute equals 60 seconds;   50′ x 60″ = 3,000″

Explanation. To change minutes to seconds, multiply the number of minutes (50′) by 60. The answer is 3,000″. 20° 50′ = 20° 3,000″.

To change seconds to minutes, divide the number of seconds by 60.

Illustration. Change 180″ to minutes.

Solution. 180″ ÷ 60″ = 3′

Explanation. Divide 180″ by 60. The answer is 3′.

LAYING OUT AND MEASURING ANGLES

To lay out an angle with a protractor first draw a line for one side of the angle. The center of the protractor should touch the point that forms the vertex of the angle. Find the size of the angle needed on the protractor. Place a dot at this point. Connect the point with the vertex of the angle.

To measure an angle with a protractor, position the protractor so that its center point meets the vertex of the angle, and one side of the angle rests along the 0° point of the protractor. The other side of the angle will rest along 2 numbers in line on the protractor. If the angle is more than 90°, figure 43-5, read the larger number to obtain the size of the angle. If it is less than 90°, figure 43-6, read the smaller number to obtain the size.

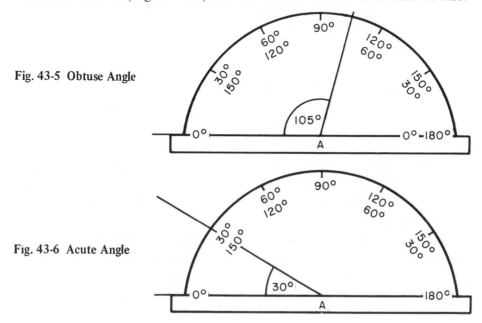

Fig. 43-5 Obtuse Angle

Fig. 43-6 Acute Angle

HOW TO READ A BEVEL PROTRACTOR

The bevel protractor with the vernier scale in figure 43-7 can be accurately read to 5 minutes (5′ or 1/12 of a degree). A *vernier scale* is made to slide along and measure parts of divisions of a larger instrument to which it is attached. The dial of the protractor is graduated both to the right and left of the zero up to 90 degrees. The vernier scale is also graduated to the right and left of zero up to 60 minutes, (60′) with each of the 12 vernier graduations representing 5 minutes. Since both the protractor dial and vernier scale have graduations in opposite directions from zero, any size angle can be measured. It should be remembered that the vernier reading must be read in the same direction from zero as the protractor, either to the left or right.

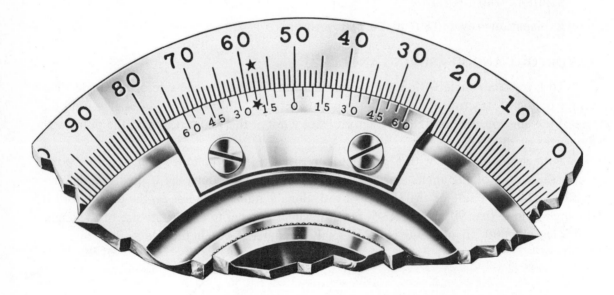

Fig. 43-7 Bevel Protractor With Vernier Scale

Each vernier graduation is 1/12 degree or 5 minutes. Therefore, if the zero graduation on the Vernier scale coincides with a graduation on the protractor dial, the reading is in exact degrees; but if some other graduation on the Vernier scale coincides with a protractor graduation, the number of Vernier graduations multiplied by 5 minutes must be added to the number of degrees read between the zeros on the protractor dial and Vernier scale.

The zero on the vernier scale in figure 43-7 is between the readings of 50 and 51 on the protractor dial to the left of the zero, indicating 50 whole degrees. Also reading to the left, the 4th line on the Vernier scale coincides with the 58 graduation on the protractor dial as indicated by the stars; therefore, 20 minutes are to be added to the number of degrees. The reading of the protractor is 50 degrees and 20 minutes (50° 20′).

ADDITION OF ANGLES

To add angles of two or more denominations, arrange the units in proper columns and add each column separately. Reduce when necessary.

Illustration. Two angles measure 15° 56′ 55″ and 10° 36′ 25″. Find the sum of the angles.

Solution. 15° 56′ 55″
$\underline{+\ 10°\ 36′\ 25″}$
25° 92′ 80″

80″ = 1′ 20″
92′ = 1° 32′

Answer = 26° 33′ 20″

Explanation. Begin at the right column. • Add 55″ and 25″ which equal 80″. There are 1′ and 20″ in 80″. • Write 20″ under the 80″. • Add the 1′ to 92″ which equals 93″. There are 1° 33′ in 93′. Write 33′ and add 1° to 25°. The sum is 26° 33′ 20″.

SUBTRACTION OF ANGLES

To subtract angles of two or more denominations, arrange the units in proper columns and subtract each column separately. Begin at the right. Borrow when necessary.

Illustration. Subtract 52° 25′ 15″ from 73° 20′ 10″.

Solution. 73° 20′ 10″
$\underline{-\ 52°\ 25′\ 15″}$
20° 54′ 55″

Explanation. Each number in the minuend must be as large or larger than the corresponding number in the subtrahend before subtracting. • Borrow 1′ from 20′ which leaves 19′. The table shows 1′ equals 60″. • Add 60″ to 10″ which equals 70″. • Subtract 15″ from 70″, leaving 55″. • Write 55″ under 15″ in the second column. • Subtract the minutes column. • Borrow 1° from 73°, leaving 72. The table shows 1° equals 60′. • Add 60′ to 19′ equals 79′. • Subtract 25′ from 79′ which leaves 54′. • Subtract in the degrees column 52° from 72° which equals 20°. The answer is 20° 54′ 55″.

MULTIPLICATION OF ANGLES

To multiply angles of different units, multiply each unit by the multiplier. Reduce the product to proper units.

Illustration. Multiply 5° 25′ 20″ by 4.

Solution. 5° 25′ 20″
$\underline{\qquad\qquad \times\ 4}$
20° 100′ 80″

80″ = 1′ 20″
100′ + 1′ = 101′
101′ = 1° 41′
20° + 1° = 21°

Answer = 21° 41′ 20″

Explanation. • Arrange the degrees, minutes, and seconds for multiplication in the same manner as for whole numbers. • Place the 4 under the last number in the smallest unit and multiply each unit separately. In this case, 4 times 20″ equals 80″. • Write 80″ in the seconds column. • Next, 4 times 25′ equals 100′ minutes. • Write 100′ under the minutes. • Multiply 5° by 4 which equals 20°. Change 80″ to 1′ 20″. • Write 20″ in the proper column. • Add the 1′ to 100′ which equals 101′; 101′ equals 1° 41′. • Write 41′ in the minutes column, cancel 100′, and add 1° to 20°. The product is 21° 41′ 20″.

DIVISION OF ANGLES

To divide angles which contain different units, divide each unit separately by the divisor. When a unit is not an exact divisor, reduce the remainder to the next smaller unit and add with the given number of smaller units to form the next partial dividend.

Illustration. Divide 21° 41′ 20″ by 4.

$$
\begin{array}{l}
\phantom{\text{Solution. 4)}} \underline{5°\ 25'\ 20''} \\
\text{Solution. 4)}\ 21°\ 41'\ 20'' \\
\phantom{\text{Solution. 4)}}\ \underline{20} \\
\phantom{\text{Solution. 4)}}\ 1° = 60' \\
\phantom{\text{Solution. 4)}}\quad\ 60' + 41' = 101' \\
\phantom{\text{Solution. 4)}}\qquad\quad\ \underline{100} \\
\phantom{\text{Solution. 4)}}\qquad\qquad\ 1' = 60'' \\
\phantom{\text{Solution. 4)}}\qquad\qquad\ 60'' + 20'' = 80'' \\
\phantom{\text{Solution. 4)}}\qquad\qquad\qquad\quad\ \underline{80}
\end{array}
$$

Answer = 5° 25′ 20″.

Explanation. Place the angle measurement under the long division sign. Begin with the units in degrees. In this case, 4 goes into 21° about 5 times. • Write 5° in the quotient over degrees. • Multiply 5 times 4 which equals 20. Subtract 20 from 21, leaving 1°. • There are 60′ in 1°, so add 60 to 41′ which equals 101′. • Divide 4 into 101. • Write 25′ in the quotient over 41′ and multiply 25 times 4 which equals 100. • Subtract 100 from 101, leaving 1′. • Change 1′ to 60″ and add 60″ to 20″, which equals 80″. • Divide 4 into 80 which equals 20. • Write 20″ in the quotient over 20″ and multiply 20 times 4 equals 80. The answer is 5° 25′ 20″.

APPLICATION

1. Change degrees to minutes in each of the following.

   a. 5°   c. 8°   e. 22°   g. 75°   i. 7° 22′
   b. 7°   d. 20°   f. 13°   h. 5° 30′ j. 32° 8′

2. Change minutes to degrees in each of the following. Express the remainder in minutes.

   a. 120′   c. 240′   e. 360′   g. 90′   i. 390′
   b. 180′   d. 300′   f. 30′   h. 150′   j. 900′

3. Change each of the following measurements to seconds.

   a. 29'   c. 50'   e. 35'   g. 7' 7"   i. 37' 37"
   b. 19'   d. 52'   f. 15' 7"   h. 58' 30"   j. 12' 10"

4. Change seconds to minutes in each of the following. Express the remainder in seconds.

   a. 3,000"   c. 420"   e. 5,400"   g. 145"   i. 125"
   b. 2940"   d. 3,300"   f. 95"   h. 99"   j. 135"

5. Use a protractor to construct the following angles.

   a. 90°   c. 30°   e. 75°   g. 20°   i. 17°
   b. 60°   d. 15°   f. 35°   h. 25°   j. 40°

6. Add the following angles.

   a. 32° 31' 40"   c. 45° 57' 19"
      27° 18' 32"   32° 30' 15"
   + 20° 30' 5"   + 42° 45' 14"

   b. 45° 55' 45"   d. 10° 40' 59"
      50° 52' 20"   12° 45' 55"
   + 37° 45' 15"   +   55' 15"

7. Subtract the following angles.

   a. 32° 31' 40"   c. 16° 43' 30"
   − 27° 18' 32"   − 12° 53' 40"

   b. 32° 31' 40"   d. 34° 17' 58"
   − 27° 45' 55"   − 31° 56' 59"

8. Multiply the following angles.

   a. 28° 32'   c. 18° 43' 35"
        x 5   x   6

   b. 32° 45'   d. 21° 0' 42"
        x 8   x   7

9. Divide the following angles. Express any remainders in seconds.

   a. 8) 32° 45'   c. 6) 45° 40' 25"

   b. 5) 68° 15' 20"   d. 3) 21° 33' 27"

CAREER PROFILE: THE SUPERVISOR

Job Description

Supervisors are hired by contractors to prepare, organize, and direct the work effort on a job site. Supervisors are responsible for the accomplishment of various duties which are involved in a construction job, and must coordinate these activities. These persons must be constantly aware of working conditions, safety regulations, and costs of the project.

Qualifications

Supervisors must be aware of all technical aspects of carpentry. They usually receive training in many fields, such as quality and waste control, planning, methods of construction, and safety. Some may take additional courses in management, industrial organization, and economics. As with other professions in carpentry, an apprenticeship is usually the best place to begin learning the duties and practices of a supervisor.

Supervising

Section 9 Fundamentals of Trigonometry

Unit 44
Right Angle Trigonometry

OBJECTIVES

After studying this unit, the student should be able to

- label the sides of a right triangle as hypotenuse, opposite side, and adjacent side.
- read sine, cosine, and tangent in a table of trigonometric functions.

Trigonometry is a branch of mathematics in which unknown angles and sides of triangles are determined. In trigonometry, sides of right triangles are identified as the hypotenuse, the opposite side, or the adjacent side. The longest side of the right triangle, which is always the side opposite the right angle, is the *hypotenuse.* The position of the opposite side and the adjacent side depends upon the reference angle. The *opposite side* is always opposite the angle used as a reference. The *adjacent side* is always next to the reference angle.

Consider the right triangles in figure 44-1. When angle x is used as the reference angle, side b is the adjacent side and side a is the opposite side. When angle y is the reference angle, side b is the opposite side and side a is the adjacent side.

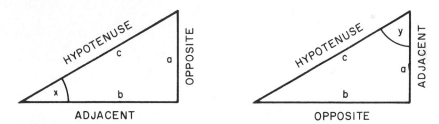

Fig. 44-1 Sides of a Right Triangle

In trigonometry, the three vertices (plural of vertex) are identified with capital letters of the alphabet; the side opposite a specific vertex is identified with the lower case of the same letter, figure 44-2.

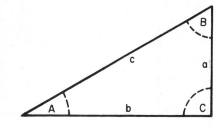

Fig. 44-2 Trigonometric Functions of a Right Triangle

TRIGONOMETRIC FUNCTIONS

There are six *functions* in trigonometry, or ratios between sides of triangles. For the scope of this text, it is necessary to discuss three of these functions: the sine, cosine, and tangent. Each of these functions describes a ratio between two specific sides of a triangle, figure 44-3. These three functions are discussed separately in the following units.

$$\frac{\text{opposite side}}{\text{hypotenuse}} \quad \frac{\text{opp}}{\text{hyp}} \quad \frac{a}{c} = \text{sine of angle A} = \sin A$$

$$\frac{\text{adjacent side}}{\text{hypotenuse}} \quad \frac{\text{adj}}{\text{hyp}} = \frac{b}{c} = \text{cosine of angle A} = \cos A$$

$$\frac{\text{opposite side}}{\text{adjacent side}} \quad \frac{\text{opp}}{\text{adj}} = \frac{a}{b} = \text{tangent of angle A} = \tan A$$

Fig. 44-3 Trigonometric Functions with Abbreviations

The size of the angles is determined by the ratio of two of the sides of the triangles, and the ratio of the two sides determines the size of the angle. The values of the ratios change with every change in the size of an angle. The ratio between side a (opposite side) and side c (hypotenuse) is the sine of ∠ A.

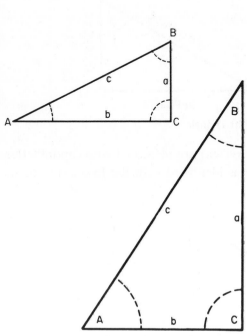

Fig. 44-4 As ∠ A Increases, Sides a and c Increase

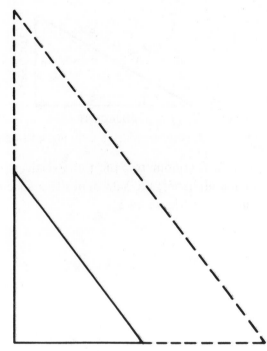

Fig. 44-5 Increase in Length of Sides Does Not Change Size of Angle

Angles	Sin	Cos	Tan	Cot	Angles
0° 00'	.00000	1.0000	.00000	infinite	90° 00'
10'	.00291	1.0000	.00291	343.77	50'
20'	.00582	.99998	.00582	171.88	40'
30'	.00873	.99996	.00873	114.59	30'
40'	.01164	.99993	.01164	85.940	20'
50'	.01454	.99989	.01455	68.750	10'
1° 00'	.01745	.99985	.01746	57.290	89° 00'
10'	.02036	.99979	.02036	49.104	50'
20'	.02327	.99973	.02328	42.964	40'
30'	.02618	.99966	.02619	38.189	30'
40'	.02908	.99958	.02910	34.368	20'
50'	.03199	.99949	.03201	31.242	10'
2° 00'	.03490	.99939	.03492	28.636	88° 00'
10'	.03781	.99929	.03783	26.432	50'
20'	.04071	.99917	.04075	24.542	40'
30'	.04362	.99905	.04366	22.904	30'
40'	.04653	.99892	.04658	21.470	20'
50'	.04943	.99878	.04949	20.206	10'
3° 00'	.05234	.99863	.05241	19.081	87° 00'
10'	.05524	.99847	.05533	18.075	50'
20'	.05814	.99831	.05824	17.169	40'
30'	.06105	.99813	.06116	16.350	30'
40'	.06395	.99795	.06408	15.605	20'
50'	.06685	.99776	.06700	14.924	10'
4° 00'	.06976	.99756	.06993	14.301	86° 00'
10'	.07266	.99736	.07285	13.727	50'
20'	.07556	.99714	.07578	13.197	40'
30'	.07846	.99692	.07870	12.706	30'
40'	.08136	.99668	.08163	12.251	20'
50'	.08426	.99644	.08456	11.826	10'
Angles	Cos	Sin	Cot	Tan	Angles

Angles	Sin	Cos	Tan	Cot	Angles
5° 00'	.08716	.99619	.08749	11.430	85° 00'
10'	.09005	.99594	.09042	11.059	50'
20'	.09295	.99567	.09335	10.712	40'
30'	.09585	.99540	.09629	10.385	30'
40'	.09874	.99511	.09923	10.078	20'
50'	.10164	.99482	.10216	9.7882	10'
6° 00'	.10453	.99452	.10510	9.5144	84° 00'
10'	.10742	.99421	.10805	9.2553	50'
20'	.11031	.99390	.11099	9.0098	40'
30'	.11320	.99357	.11394	8.7769	30'
40'	.11609	.99324	.11688	8.5556	20'
50'	.11898	.99290	.11983	8.3450	10'
7° 00'	.12187	.99255	.12278	8.1444	83° 00'
10'	.12476	.99219	.12574	7.9530	50'
20'	.12764	.99182	.12869	7.7704	40'
30'	.13053	.99144	.13165	7.5958	30'
40'	.13341	.99106	.13461	7.4287	20'
50'	.13629	.99067	.13758	7.2687	10'
8° 00'	.13917	.99027	.14054	7.1154	82° 00'
10'	.14205	.98986	.14351	6.9682	50'
20'	.14493	.98944	.14648	6.8269	40'
30'	.14781	.98902	.14945	6.6912	30'
40'	.15069	.98858	.15243	6.5606	20'
50'	.15356	.98814	.15540	6.4348	10'
9° 00'	.15643	.98769	.15838	6.3138	81° 00'
10'	.15931	.98723	.16137	6.1970	50'
20'	.16218	.98676	.16435	6.0844	40'
30'	.16505	.98629	.16734	5.9758	30'
40'	.16792	.98580	.17033	5.8708	20'
50'	.17078	.98531	.17333	5.7694	10'
Angles	Cos	Sin	Cot	Tan	Angles

Angles	Sin	Cos	Tan	Cot	Angles
10° 00'	.17365	.98481	.17633	5.6713	80° 00'
10'	.17651	.98430	.17933	5.5764	50'
20'	.17937	.98378	.18233	5.4845	40'
30'	.18224	.98325	.18534	5.3955	30'
40'	.18509	.98272	.18835	5.3093	20'
50'	.18795	.98218	.19136	5.2257	10'
11° 00'	.19081	.98163	.19438	5.1446	79° 00'
10'	.19366	.98107	.19740	5.0658	50'
20'	.19652	.98050	.20042	4.9894	40'
30'	.19937	.97992	.20345	4.9152	30'
40'	.20222	.97934	.20648	4.8430	20'
50'	.20507	.97875	.20952	4.7729	10'
12° 00'	.20791	.97815	.21256	4.7046	78° 00'
10'	.21076	.97754	.21560	4.6383	50'
20'	.21360	.97692	.21864	4.5736	40'
30'	.21644	.97630	.22169	4.5107	30'
40'	.21928	.97566	.22475	4.4494	20'
50'	.22212	.97502	.22781	4.3897	10'
13° 00'	.22495	.97437	.23087	4.3315	77° 00'
10'	.22778	.97371	.23393	4.2747	50'
20'	.23062	.97304	.23700	4.2193	40'
30'	.23345	.97237	.24008	4.1653	30'
40'	.23627	.97169	.24316	4.1126	20'
50'	.23910	.97100	.24624	4.0611	10'
14° 00'	.24192	.97030	.24933	4.0108	76° 00'
10'	.24474	.96959	.25242	3.9617	50'
20'	.24756	.96887	.25552	3.9136	40'
30'	.25038	.96815	.25862	3.8667	30'
40'	.25320	.96742	.26172	3.8208	20'
50'	.25601	.96667	.26483	3.7760	10'
Angles	Cos	Sin	Cot	Tan	Angles

Fig. 44-6 Trigonometric Functions: Sine, Cosine, Tangent

Angles	Cot	Tan	Cos	Sin	Angles
25° 00'	2.1445	.46631	.90631	.42262	65° 00'
10'	2.1283	.46985	.90507	.42525	50'
20'	2.1123	.47341	.90383	.42788	40'
30'	2.0965	.47698	.90259	.43051	30'
40'	2.0809	.48055	.90133	.43313	20'
50'	2.0655	.48414	.90007	.43575	10'
26° 00'	2.0503	.48773	.89879	.43837	64° 00'
10'	2.0353	.49134	.89752	.44098	50'
20'	2.0204	.49495	.89623	.44359	40'
30'	2.0057	.49858	.89493	.44620	30'
40'	1.9912	.50222	.89363	.44880	20'
50'	1.9768	.50587	.89232	.45140	10'
27° 00'	1.9626	.50953	.89101	.45399	63° 00'
10'	1.9486	.51319	.88968	.45658	50'
20'	1.9347	.51688	.88835	.45917	40'
30'	1.9210	.52057	.88701	.46175	30'
40'	1.9074	.52427	.88566	.46433	20'
50'	1.8940	.52798	.88431	.46690	10'
28° 00'	1.8807	.53171	.88295	.46947	62° 00'
10'	1.8676	.53545	.88158	.47204	50'
20'	1.8546	.53920	.88020	.47460	40'
30'	1.8418	.54296	.87882	.47716	30'
40'	1.8291	.54673	.87743	.47971	20'
50'	1.8165	.55051	.87603	.48226	10'
29° 00'	1.8041	.55431	.87462	.48481	61° 00'
10'	1.7917	.55812	.87321	.48735	50'
20'	1.7796	.56194	.87178	.48989	40'
30'	1.7675	.56577	.87036	.49242	30'
40'	1.7556	.56962	.86892	.49495	20'
50'	1.7438	.57348	.86748	.49748	10'
Angles	Tan	Cot	Sin	Cos	Angles

Angles	Cot	Tan	Cos	Sin	Angles
20° 00'	2.7475	.36397	.93969	.34202	70° 00'
10'	2.7228	.36727	.93869	.34475	50'
20'	2.6985	.37057	.93769	.34748	40'
30'	2.6746	.37388	.93667	.35021	30'
40'	2.6511	.37720	.93565	.35293	20'
50'	2.6279	.38053	.93462	.35565	10'
21° 00'	2.6051	.38386	.93358	.35837	69° 00'
10'	2.5826	.38721	.93253	.36108	50'
20'	2.5605	.39055	.93148	.36379	40'
30'	2.5387	.39391	.93042	.36650	30'
40'	2.5172	.39727	.92935	.36921	20'
50'	2.4960	.40065	.92827	.37191	10'
22° 00'	2.4751	.40403	.92718	.37461	68° 00'
10'	2.4545	.40741	.92609	.37730	50'
20'	2.4342	.41081	.92499	.37999	40'
30'	2.4142	.41421	.92388	.38268	30'
40'	2.3945	.41763	.92276	.38537	20'
50'	2.3750	.42105	.92164	.38805	10'
23° 00'	2.3559	.42447	.92050	.39073	67° 00'
10'	2.3369	.42791	.91936	.39341	50'
20'	2.3183	.43136	.91822	.39608	40'
30'	2.2998	.43481	.91706	.39875	30'
40'	2.2817	.43828	.91590	.40141	20'
50'	2.2637	.44175	.91472	.40408	10'
24° 00'	2.2460	.44523	.91355	.40674	66° 00'
10'	2.2286	.44872	.91236	.40939	50'
20'	2.2113	.45222	.91116	.41204	40'
30'	2.1943	.45573	.90996	.41469	30'
40'	2.1775	.45924	.90875	.41734	20'
50'	2.1609	.46277	.90753	.41998	10'
Angles	Tan	Cot	Sin	Cos	Angles

Angles	Cot	Tan	Cos	Sin	Angles
15° 00'	3.7321	.26795	.96593	.25882	75° 00'
50'	3.6891	.27107	.96517	.26163	50'
40'	3.6471	.27419	.96440	.26443	40'
30'	3.6059	.27732	.96363	.26724	30'
20'	3.5656	.28046	.96285	.27004	20'
10'	3.5261	.28360	.96206	.27284	10'
16° 00'	3.4874	.28675	.96126	.27564	74° 00'
50'	3.4495	.28990	.96046	.27843	50'
40'	3.4124	.29305	.95964	.28123	40'
30'	3.3759	.29621	.95882	.28402	30'
20'	3.3402	.29938	.95799	.28680	20'
10'	3.3052	.30255	.95715	.28959	10'
17° 00'	3.2709	.30573	.95630	.29237	73° 00'
50'	3.2371	.30891	.95545	.29515	50'
40'	3.2041	.31210	.95459	.29793	40'
30'	3.1716	.31530	.95372	.30071	30'
20'	3.1397	.31850	.95284	.30348	20'
10'	3.1084	.32171	.95195	.30625	10'
18° 00'	3.0777	.32492	.95106	.30902	72° 00'
50'	3.0475	.32814	.95015	.31178	50'
40'	3.0178	.33136	.94924	.31454	40'
30'	2.9887	.33460	.94832	.31730	30'
20'	2.9600	.33783	.94740	.32006	20'
10'	2.9319	.34108	.94646	.32282	10'
19° 00'	2.9042	.34433	.94552	.32557	71° 00'
50'	2.8770	.34758	.94457	.32832	50'
40'	2.8502	.35085	.94361	.33106	40'
30'	2.8239	.35412	.94264	.33381	30'
20'	2.7980	.35740	.94167	.33655	20'
10'	2.7725	.36068	.94068	.33929	10'
Angles	Sin	Cos	Cot	Tan	Angles

Fig. 44-6 Trigonometric Functions: Sine, Cosine, Tangent (cont'd)

Angles	Sin	Cos	Tan	Cot	Angles
30° 00'	.50000	.86603	.57735	1.7321	60° 00'
10'	.50252	.86457	.58124	1.7205	50'
20'	.50503	.86310	.58513	1.7090	40'
30'	.50754	.86163	.58904	1.6977	30'
40'	.51004	.86015	.59297	1.6864	20'
50'	.51254	.85866	.59691	1.6753	10'
31° 00'	.51504	.85717	.60086	1.6643	59° 00'
10'	.51753	.85567	.60483	1.6534	50'
20'	.52002	.85416	.60881	1.6426	40'
30'	.52250	.85264	.61280	1.6319	30'
40'	.52498	.85112	.61681	1.6213	20'
50'	.52745	.84959	.62083	1.6107	10'
32° 00'	.52992	.84805	.62487	1.6003	58° 00'
10'	.53238	.84650	.62892	1.5900	50'
20'	.53484	.84495	.63299	1.5798	40'
30'	.53730	.84339	.63707	1.5697	30'
40'	.53975	.84182	.64117	1.5597	20'
50'	.54220	.84025	.64528	1.5497	10'
33° 00'	.54464	.83867	.64941	1.5399	57° 00'
10'	.54708	.83708	.65355	1.5301	50'
20'	.54951	.83549	.65771	1.5204	40'
30'	.55194	.83389	.66189	1.5108	30'
40'	.55436	.83228	.66608	1.5013	20'
50'	.55678	.83066	.67028	1.4919	10'
34° 00'	.55919	.82904	.67451	1.4826	56° 00'
10'	.56160	.82741	.67875	1.4733	50'
20'	.56401	.82577	.68301	1.4641	40'
30'	.56641	.82413	.68728	1.4550	30'
40'	.56880	.82248	.69157	1.4460	20'
50'	.57119	.82082	.69588	1.4370	10'
Angles	Cos	Sin	Cot	Tan	Angles

Angles	Cot	Tan	Cos	Sin	Angles
55° 00'	1.4282	.70021	.81915	.57358	35° 00'
50'	1.4193	.70455	.81748	.57596	10'
40'	1.4106	.70891	.81580	.57833	20'
30'	1.4020	.71329	.81412	.58070	30'
20'	1.3934	.71769	.81242	.58307	40'
10'	1.3848	.72211	.81072	.58543	50'
54° 00'	1.3764	.72654	.80902	.58779	36° 00'
50'	1.3680	.73100	.80730	.59014	10'
40'	1.3597	.73547	.80558	.59248	20'
30'	1.3514	.73996	.80386	.59482	30'
20'	1.3432	.74447	.80212	.59716	40'
10'	1.3351	.74900	.80038	.59949	50'
53° 00'	1.3270	.75355	.79864	.60182	37° 00'
50'	1.3190	.75812	.79688	.60414	10'
40'	1.3111	.76272	.79512	.60645	20'
30'	1.3032	.76733	.79335	.60876	30'
20'	1.2954	.77196	.79158	.61107	40'
10'	1.2876	.77661	.78980	.61337	50'
52° 00'	1.2799	.78129	.78801	.61566	38° 00'
10'	1.2723	.78598	.78622	.61795	10'
20'	1.2647	.79070	.78442	.62024	20'
30'	1.2572	.79544	.78261	.62251	30'
40'	1.2497	.80020	.78079	.62479	40'
50'	1.2423	.80498	.77897	.62706	50'
51° 00'	1.2349	.80978	.77715	.62932	39° 00'
50'	1.2276	.81461	.77531	.63158	10'
40'	1.2203	.81946	.77347	.63383	20'
30'	1.2131	.82434	.77162	.63608	30'
20'	1.2059	.82923	.76977	.63832	40'
10'	1.1988	.83415	.76791	.64056	50'
Angles	Tan	Cot	Sin	Cos	Angles

Angles	Cot	Tan	Cos	Sin	Angles
50° 00'	1.1918	.83910	.76604	.64279	40° 00'
50'	1.1847	.84407	.76417	.64501	10'
40'	1.1778	.84906	.76229	.64723	20'
30'	1.1709	.85408	.76041	.64945	30'
20'	1.1640	.85912	.75851	.65166	40'
10'	1.1572	.86419	.75661	.65386	50'
49° 00'	1.1504	.86929	.75471	.65606	41° 00'
50'	1.1436	.87441	.75280	.65825	50'
40'	1.1369	.87955	.75088	.66044	40'
30'	1.1303	.88473	.74896	.66262	30'
20'	1.1237	.88992	.74703	.66480	20'
10'	1.1171	.89515	.74509	.66697	10'
48° 00'	1.1106	.90040	.74314	.66913	42° 00'
50'	1.1041	.90569	.74120	.67129	50'
40'	1.0977	.91099	.73924	.67344	40'
30'	1.0913	.91633	.73728	.67559	30'
20'	1.0850	.92170	.73531	.67773	20'
10'	1.0786	.92709	.73333	.67987	10'
47° 00'	1.0724	.93252	.73135	.68200	43° 00'
50'	1.0661	.93797	.72937	.68412	10'
40'	1.0599	.94345	.72737	.68624	20'
30'	1.0538	.94896	.72537	.68835	30'
20'	1.0477	.95451	.72337	.69046	40'
10'	1.0416	.96008	.72136	.69256	50'
46° 00'	1.0355	.96569	.71934	.69466	44° 00'
50'	1.0295	.97132	.71732	.69675	50'
40'	1.0236	.97700	.71529	.69883	40'
30'	1.0176	.98270	.71325	.70091	30'
20'	1.0117	.98843	.71121	.70298	20'
10'	1.0058	.99420	.70916	.70505	10'
45° 00'	1.0000	1.0000	.70711	.70711	45° 00'
Angles	Tan	Cot	Sin	Cos	Angles

Fig. 44-6 Trigonometric Functions: Sine, Cosine, Tangent (cont'd)

READING THE TABLE OF TRIGONOMETRIC FUNCTIONS

Figure 44-6 gives the decimal equivalents of the ratios of two sides of a triangle for the three functions that this text discusses. The functions are listed for angles measuring from 0° to 90° in 10′ steps.

Angles from 0°00′ to 45°00′ are listed on the left side of the table under the word "angle". The function names to be used for these angles are at the top of the table. Angles measuring from 45°00′ to 90°00′, which appear on the right side of the table, have function names at the bottom.

To locate a function of an angle measuring from 0°00′ to 45°00′, look for the angle measurement in the left column and locate the proper function at the top of the table. Move down the function row until the desired angle reading is reached. To locate the appropriate minutes, read from the top of the chart to the bottom. Record the answer.

Illustration. Find the sine of 6°40′.

Solution. sin 6°40′ = .11609

Note: When giving the actual measurement, use the abbreviation "sin" for "sine."

Explanation. In figure 44-6, find 6°00′; under 6°00′, find 40′. Find "sin" at the top of the table and follow down this column until the number .11609 is reached. Record the answer.

To find the functions of an angle above 45°00′ look for the degrees and minutes in the right column of the table above the word "angle" and look for the function at the bottom of the page. Go up the function column until the desired degree is reached and then record it. To locate minutes when the angle is greater than 45°, read from the bottom of the chart to the top.

Illustration. Find the cosine of 62°00′ in figure 44-6. Above the word "angle" at the right side of the page find 62°00′. Read at the bottom of the table the abbreviation "cos". In the cosine column opposite 62°00′ is the cosine function .46947. The cosine of 62°00′ is .46947.

Illustration. Find the tangent of 60°10′ in figure 44-6. Below the word "angle" at the right side of the page is 60°10′. Find the word "tan" at the bottom of the table. In the tangent column opposite 60°10′ is the tangent function 1.7438. The tangent function of 60°10′ = 1.7438.

APPLICATION

1. Define sine.

2. Define tangent.

3. Define cosine.

4. Make a drawing of a right triangle and label each side.

5. Using figure 44-6, find the sines for the following angles.

a. sin 30°	c. sin 19°	e. sin 70°	g. sin 65°	i. sin 90°
b. sin 20°	d. sin 45°	f. sin 55°	h. sin 81°	j. sin 89°

6. Using figure 44-6, find the tangents for the following angles.

    a. tan 40°   f. tan 55°
    b. tan 42°   g. tan 57°
    c. tan 18°   h. tan 71°
    d. tan 35°   i. tan 63°
    e. tan 44°   j. tan 11°

7. Using table 44-6, find the cosines for the following angles.

    a. cos 16°   f. cos 42°30′
    b. cos 50°   g. cos 61°10′
    c. cos 72°   h. cos 36°20′
    d. cos 32°   i. cos 40°40′
    e. cos 28°   j. cos 58°20′

Unit 45
The Sine Function

OBJECTIVES

After studying this unit, the student should be able to

- find the value of sin A and sin B in a right triangle.
- find the height of side a in a right triangle.
- find the area of a parallelogram and trapezoid by using the sine formula.
- apply this information to specific problems in carpentry.

The *sine* of an angle is the ratio of the side opposite the angle to the hypotenuse, figure 45-1. It is stated as sin = opposite side/hypotenuse and the formula is sin = a/c.

By definition, an angle that is less than 90° is an acute angle. The ratio of the length of the side opposite an acute angle to the length of the hypotenuse is the sine of the acute angle.

Illustration. The right triangle (ABC) in figure 45-2 has a hypotenuse of 10″; the side opposite ∠A is 6″. Find sin A and sin B.

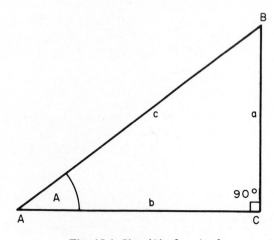

Fig. 45-1 Sine (A) of an Angle

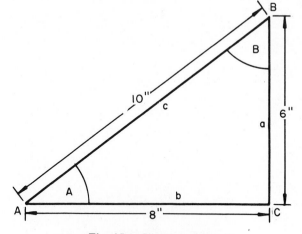

Fig. 45-2 Sine A and Sine B

Solution. 1. sin A = a/c

  sin A = 6″/10″ = 6 ÷ 10 = .6

  sin A = .6

  2. sin b = side opposite B/hypotenuse

  sin B = b/c

  sin B = 8/10 = 8 ÷ 10 = .8

  sin B = .8

266

Explanation. (1) The side opposite angle A measures 6″. The hypotenuse is 10″. The formula is sin A = a/c. Substitute 6″ for a and 10″ for c in the formula. Divide 6 by 10. The result is .6. Sine A = .6. (2) The side opposite angle B is 8″. The hypotenuse is 10″. The formula is sin B = b/c. Substitute 8″ for b and 10″ for c in the formula. Divide 8 by 10. The result is .8.

PRACTICAL APPLICATION

In carpentry, the sine formula can be used to find the height of a rafter from the plate to the ridge if the length of the rafter and the number of degrees in the angle at the plate are known.

Illustration. The length of the rafter in figure 45-3 is 10′. The angle at the plate is 30°. What is the height of the ridge above the plate?

Solution. Use the formula a/c = sin A.

Step 1.   x/r = sin 30°

Step 2.   x/10 = sin 30°

Step 3.   x = 10 (.5000)

Step 4.   x = 5.0000

x = 5′

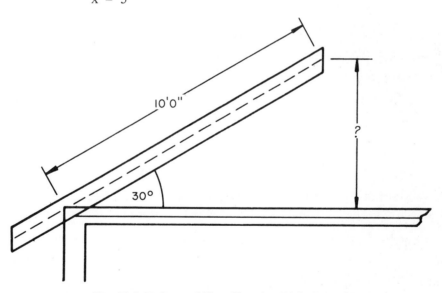

Fig. 45-3 Rafter and Plate Forming Right Triangle.

Explanation. The rafter is in the position of a hypotenuse in a right triangle. The angle at the plate is in the position of the angle A in the triangle. • Use the formula x/r = sin 30°. • Figure 44-6 shows that the value of sin 30° is .5000. • Write 10 in the place of r and .5000 in place of sin 30°. • The x is unknown; x is 10 times .5000, which equals 5′.

The measurement of the angle at the plate where the rafter rests can be found if the rafter length and the rise are known.

Illustration. In figure 45-4, the length of the rafter is 10′ and its rise is 5′. What is the measurement of the angle at the plate?

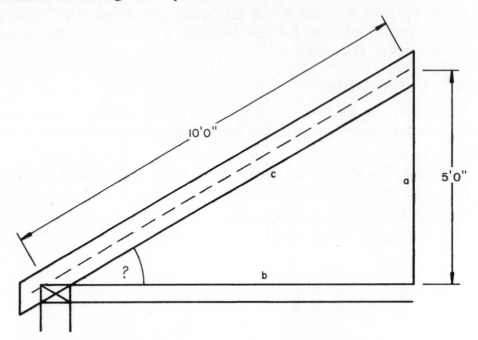

Fig. 45-4 Rafter and Plate Forming Right Triangle.

Solution. Use the formula sin A = a/c.

  sin A = opposite/hypotenuse

  Step 1.   sin A = 5/10

  Step 2.   sin A = .5

  Step 3.   A = 30°

Explanation. Use the formula sin A = opposite/hypotenuse.

• In the formula substitute 5 for a and 10 for c. • Sin A equals 5 divided by 10 which equals .5.   • The .5 equals sin 30°, which is the angle of the rafter at the plate.

The length of a rafter can be determined by the formula a/c = sin A.

Illustration. Find the length of a rafter whose rise is 7′ 6″ and the angle at the plate is sin 26° 10′. Carry out the answer to the nearest whole number.

Solution. a/c = sin A

  Step 1.   c = a/sin A

  Step 2.   c = 7.5/.44098

  Step 3.   c = 17′

Explanation. Since a (7.5) is known and sin A equals .44098, the unknown can be found by dividing 7.5 by .44098. Rounded off, the result is 17′, which represents the length of the rafter.

Treads and risers are supported by stringers in stair construction. The angle at the base makes the steps safe and comfortable for walking from one level of a structure to another. The carpenter should be able to determine the angle when the total rise and length of the stringer are known.

Illustration. The rise of the staircase in figure 45-5 is 84″ and the stringer is 132″. What is the angle at the base to the nearest 10′?

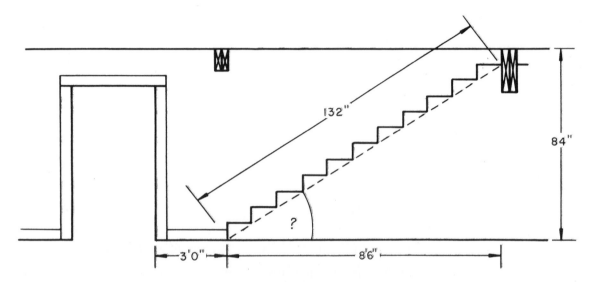

Fig. 45-5 Angle at the Base of Stairs

Solution. Use the formula sin A = a the rise/c stringer.

sin A = 84/132

sin A = .63636

sin A = .6364 (rounded off)

A = approx. 39°30′

Explanation. The unknown is the angle at the base. • The opposite side is 84″, and the hypotenuse is 132″. The formula is sin A = a/c. The rise is a in the formula, and the stringer is c. • Substitute 84 for a and 132 for c. • Divide 84 by 132, which equals .63636. In figure 45-5, the number .63636 is between .63608 and .63832. The value of .63608 is 39° 30′, and the value of .63832 is 39° 40′. The value .63636 is nearest .63608; therefore, the approximate angle measurement to the nearest 10′ is 39° 30′.

DETERMINING AREA

Area may be determined by applying the sine formula.

Illustration. Find the area of parallelogram ABCD, figure 45-6. Length AB = 35″, A = 65°, and AD = 25″ Round off the answer to square inches.

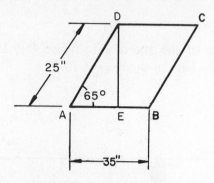

Fig. 45-6 Parallelogram

Solution.   Use the formula sin A = h/c.   h = length DE; c = length AD

Step 1.   h/c = sin A

h/25 = sin 65°

sin 65° = .90631

h = 25 x .90631

h = 22.65775

Step 2.   Area = hb.

Area = 22.65775 x 35 = 793.02125 or 793 sq. in.

Explanation. The formula for finding the area of a parallelogram is Area = hb. Since h is unknown, the sine formula is used to determine h. • The formula is written h/c = sin A. • Substitute 25 for c and sin 65° for sin A. • The value of sin 65° is .90631; h equals 25 times .90631. • The product is 22.65775 which is the distance from D to E, or the height of the parallelogram. • After h is found, use the formula Area = hb. • Multiply 22.6625 by 35, which equals 793.02125. • Rounded off, the answer is 793 sq. in.

Illustration. Find the area of the trapezoid in figure 45-7 with measurements of LM, 20″; ON, 15″; sin A = 72°; and LO, 50″. Round off your answer to square inches.

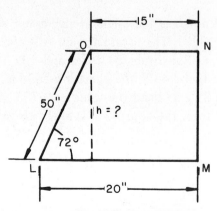

Fig. 45-7 Trapezoid

Solution. Step 1. Use the formula, h/c = sin A.

h/50 = sin 72°

h/50 = .95106

h = 50 (.95106)

h = 47.553

Step 2. Use the formula area = $\frac{t + b}{2}$ x h

Area = 17.5 x 47.553 = 832.1775=

Area = 832 sq. in.

Explanation. The formula for finding the area of a trapezoid is [(t + b) /2] x h. In this case, h is the unknown. • The formula h/c = sin A is used to find h. • Substitute 50 for c and 72° for A. • Write the formula h/50 = sin 72°.

The value of sin 72° is .95106; h/50 = .95106; h = 50 times .95106 and h equals 47.553. Then use the formula [(t + b) /2] x h to find the area. Substitute 15 for t and 20 for b in the formula. Add 15 and 20, which equals 35. Divide 35 by 2, which equals 17.5. Multiply 47.553 by 17.5 for the area, 832.1775. Rounded off, the area is 832 sq. in.

APPLICATION

1. Find the value of sin ∠ A in figure 45-8 with measurements as indicated.

a. a = 6'   c = 12'   f. a = 15"   c = 25"
b. a = 8'   c = 12'   g. a = 20"   c = 30"
c. a = 9' 6"   c = 13' 6"   h. a = 18"   c = 28"
d. a = 10'   c = 15'   i. a = 22"   c = 24"
e. a = 12' 4"   c = 18' 6"   j. a = 30"   c = 35"

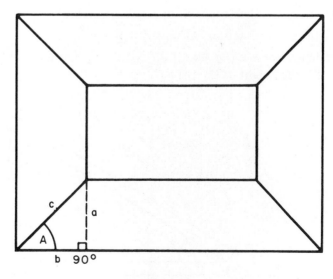

Fig. 45-8

2. Find the height of a in the trapezoid in figure 45-9 with measurements as indicated.

a. ∠ A 20°   c = 6'   f. ∠ A 68° 30'   c = 15"
b. ∠ A 40°   c = 12'   g. ∠ A 31° 10'   c = 10"
c. ∠ A 60°   c = 12'   h. ∠ A 29° 50'   c = 16"
d. ∠ A 45°   c = 8'   i. ∠ A 35° 30'   c = 32"
e. ∠ A 35°   c = 11'   j. ∠ A 40° 40'   c = 48"

Fig. 45-9

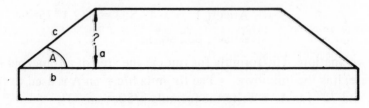

3. Find the value of sine B in figure 45-10 as indicated.

a. c = 12"   b = 8"   f. c = 10.5"   b = 8.6"
b. c = 14"   b = 8"   g. c = 11.3"   b = 9.5"
c. c = 18"   b = 12"   h. c = 12.6"   b = 10.3"
d. c = 20"   b = 14"   i. c = 14.7"   b = 10.6"
e. c = 22"   b = 20"   j. c = 15.9"   b = 12.5"

Fig. 45-10

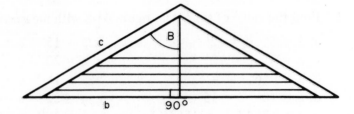

4. Determine the measurement of ∠ A, figure 45-11.

5. Determine the measurement of ∠ B, figure 45-11.

6. Determine the length of AC in figure 45-11.

7. The length of a rafter on a roof is 12' 6". The angle at the base is 30°. What is the height of the rise?

Fig. 45-11

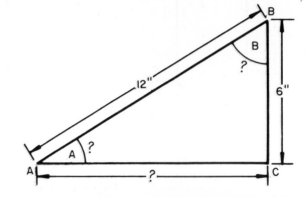

Unit 46
The Cosine Function

OBJECTIVES

After studying this unit, the student should be able to

- determine angle size in a right triangle by the cosine formula.
- find the length of sides of a right triangle by the cosine formula.
- find the length of an unknown side of a trapezoid.
- apply this information to specific problems in carpentry.

The *cosine* of an angle is the ratio of the side adjacent to that angle to the hypotenuse, figure 46-1.

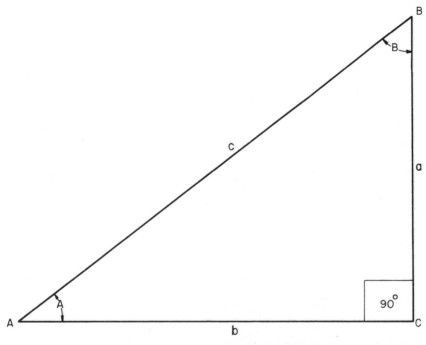

Fig. 46-1 Cosine ($\angle$A) = adjacent side/hypotenuse = b/c

The cosine formula may be used to find an unknown acute angle in a right triangle and to find the length of the unknown side. It is expressed as

$$\cos \angle A = \text{adjacent side/hypotenuse} \quad \text{or} \quad \cos \angle A = b/c.$$

DETERMINING MEASUREMENT BY THE COSINE FORMULA

Finding the Unknown Angle

If the lengths of the hypotenuse and adjacent side of a right triangle are known, the cosine of $\angle$ A may be determined.

Illustration. Find the cosine of A in figure 46-2.

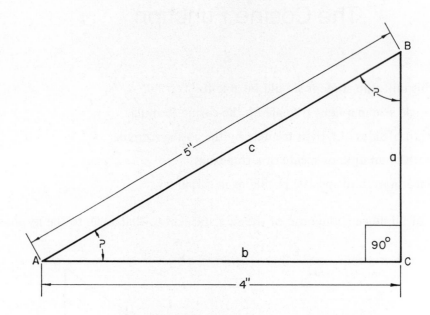

Fig. 46-2 Unknown Angles of a Right Triangle

Solution.   cos A = adjacent side/hypotenuse

cos A = b/c

cos A = 4/5

cos A = .8000

∠A = approximately 36° 50′

Explanation. • The cosine of ∠ A equals 4 divided by 5, which equals .8000. • In figure 44-6, the closest angle to cosine .8000 is 36° 50′. • For the right triangle in the figure, Cos A is approximately 36° 50′.

There are 180° in any triangle. In a right triangle, the 90° angle is a constant; the triangle also contains two acute angles. If one acute angle is known, the measurement of the other acute angle may be determined by subtraction.

Illustration. Angle A in figure 46-2 is 36° 50′. Find angle B.

Solution.   ∠B = 90° – 36° 50′

∠B = 53° 10′

Check: 53° 10′ + 36° 50′ + 90° = 180°

Explanation. • Subtract the measurement of ∠ A (36° 50′) from 90°. • The difference is 53° 10′. • When working with the cosine function, it is important to know the cosine decreases when the angle increases.

Illustration. Show that the cosine decreases when the angle increases. (1) Find cos A for figure 46-3A whose adjacent side is 2″ and hypotenuse is 3″. (2) Find Cos A for figure 46-3B whose adjacent side is 2″ and hypotenuse is 5″.

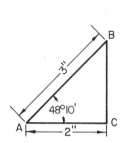

Fig. 46-3A Cosine Decreases As Angle Increases.
Cos ∠ A (48°10′) Is .6667.

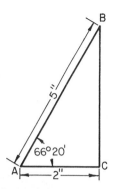

Fig. 46-3B Cosine Decreases As Angle Increases.
Cos ∠ A (66°20′) Is .4000.

Solution.   Step 1.   cos A = b/c

cos A = 2/3

cos A = .6667 (rounded off)

∠A = 48° 10′ (nearest angle)

Step 2.   cos A = b/c

cos A = 2/5

cos A = .4000

∠A = 66° 20′ (nearest angle)

.6667 − .4000 = .2667 difference

66° 20′ − 48° 10′ = 18° 10′ difference

Explanation. In the first case, cos A is .6667 and the nearest angle is 48° 10′. In the second case, cos A is .4000 and the nearest angle is 66° 20′. The difference between 48° 10′ and 66° 20′ is 18° 10′. Therefore, .6667 is .2667 larger than .4000, yet the angle for .6667 is 18° 10′ smaller. When the cosine .6667 is decreased to .4000, the angle is increased from 48° 10′ to 66° 20′.

Finding an Unknown Side of a Triangle

The cosine formula may be used to determine the length of an adjacent side of a right triangle if the length of the hypotenuse and cos A are given.

Fig. 46-4 Parts of Roof Forming a Triangle

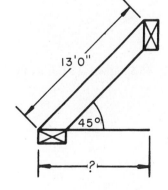

Illustration. The hypotenuse of the triangle formed by sections of the roof in figure 46-4 measures 13′ and angle A is 45°. Find the length of the adjacent side to the nearest foot.

Solution. b/c = cos A

$$b/13 = \cos 45°$$

$$\cos 45° = .70711$$

$$b = .70711 \times 13 = 9.19243′$$

$$b = \text{approx. } 9′$$

Explanation. Since the hypotenuse measures an even 13′ and angle A is 45°, the adjacent side is 13 times the cosine 45°. • The cosine 45° equals .70711. • Multiply .70711 by 13′ which equals 9.1923′. • The adjacent side is approximately 9′ long.

Note: Since angle A is 45°, angle B also measures 45° because 90° minus 45° equals 45° and 45° plus 45° plus 90° equals 180°. (There are 180° in a triangle.)

Illustration. A right triangle is formed at the corner of a structure in figure 46-5. Cosine A in the triangle is .6000 and the length of the adjacent side is 6′. Find the hypotenuse.

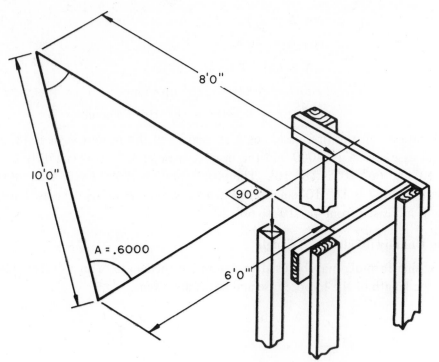

Fig. 46-5 Squaring a Corner by Trigonometric Ratio

Solution. Use the formula, cos A = b/c

$$c = b/\text{Cos } A$$

$$c = 6/.6000$$

$$c = 10′$$

Explanation. • In the formula, c is the hypotenuse. • The hypotenuse equals 6′ divided by .6000. • The answer is 10.

FINDING AN UNKNOWN SIDE OF A TRAPEZOID

The length of the top side of a trapezoid can be determined when the length of the bottom side, the length of the two slanted sides, and the two angles at the bottom of the trapezoid are known.

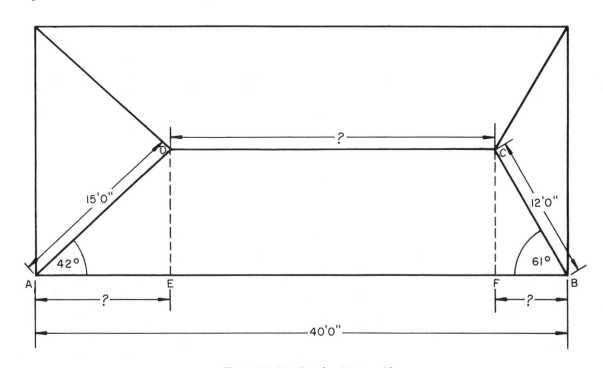

Fig. 46-6 Hip Roof – Trapezoid

Solution. Use the formula AE/AD = cos A.

Step 1.   AE/15 = cos 42°

AE = .74314 x 15 = 11.1471′

AE = 11.15′

Step 2.   BF/BC = cos B

BF/12 = cos 61°

BF/12 = .48481

BF = .48481 x 12 = 5.81772′

BF = 5.82′

Step 3.   CD = AB − (AE + BF)

CD = 40′ − 16.97′ = 23.03′

CD = 23.03′

Explanation. • To find the length of DC in the trapezoidal roof, first find the length of AE. • The angle at A is cos 42° and the length of AD is 15'. • The value of cos 42° is .74314. • The length of AE equals .74314 times 15', or 11.1471'. Rounded off, AE equals 11.15'. Then find the length of BF. • The value of cos 61° (.48481) times 12' is 5.81772. • Rounded off, BF equals 5.82'. • Add 11.15' and 5.82' which equals 16.97'. • Subtract 16.97' from 40', which equals 23.03'. • The length of DC in the trapezoidal roof is 23.03', or rounded off, 23'.

PRACTICAL APPLICATION

Finding the Angle Formed at the Base of a Rafter

If the length of a rafter and the width of a structure are known, the angle formed at the base of the rafter may be found by using the cosine formula.

Illustration. The total width of the building in figure 46-7 is 24' and the length of the common rafter is 14'. Find the nearest angle formed by the rafter and the base.

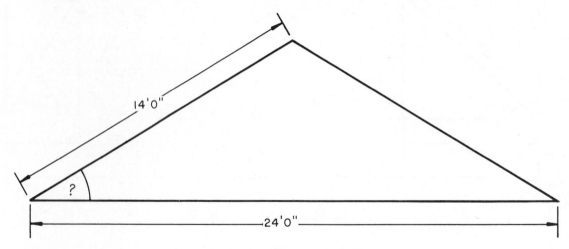

14'0"

?

24'0"

Fig. 46-7 Rafter Length and Span

Solution. 1/2 of 24' = 12'

cos A = b/c

cos A = 12/14

cos A = .85714

∠A = 31° (nearest degree)

Explanation. • First find half the width of the building: 1/2 x 24' = 12'. The 12' is b in the formula. • Divide 12' by 14' which equals .85714. • The closest value of .85714 found in figure 44-6 is 31°; therefore, the angle at the plate is approximately 31°.

Finding the Angle Formed by a Corner Brace

If the length of a corner brace and its base are known, the two acute angles formed may be found by determining angle A and subtracting angle A from 90°.

Illustration. The length of the corner brace in figure 46-8 is 10′ with a base of 6′. Find the size of the two acute angles.

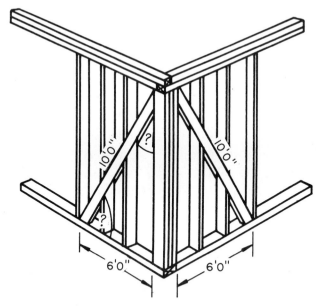

Fig. 46-8 Triangles Formed By Corner Brace

Solution. cos A = b/c

cos A = 6/10

cos A = .6000

∠A = approx. 53° 10′

∠B = 90° – 53° 10′

∠B = approx. 36° 50′

Explanation. The length of the hypotenuse and the adjacent side is known. • Divide 6′ by 10′ which equals the cosine .6000. • The closest cosine value to .6000 given in figure 44-6 is .59949 which rounded off is .5995. The measurement may be expressed as 53° 10′. Therefore, angle A equals 53° 10′. To find angle B, subtract 53° 10′ from 90°. Angle B equals approximately 36° 50′.

APPLICATION

1. Referring to figure 46-9, solve for ∠A in each of the following.

   a. c = 5″   b = 4″
   b. c = 8″   b = 6″
   c. c = 12″   b = 8″
   d. c = 14″   b = 10″
   e. c = 16″   b = 12″
   f. c = 18″   b = 14″

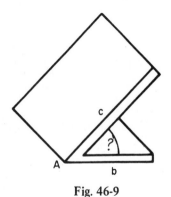

Fig. 46-9

g. c = 22"   b = 16"   k. c = 10.22"   b = 8.6"
h. c = 7.75"   b = 5.5"   l. c = 12.18"   b = 9.6"
i. c = 8.40"   b = 6.3"   m. c = 16.66"   b = 10.5"
j. c = 22"   b = 16"   n. c = 18.85"   b = 11.7"

2. Solve for b(adjacent side) in each of the following.

a. ∠ A = 22°   c = 6'   h. ∠ A = 18° 10'   c = 5.5'
b. ∠ A = 24°   c = 8'   i. ∠ A = 19° 10'   c = 7.7'
c. ∠ A = 30°   c = 10'   j. ∠ A = 20° 30'   c = 9.9'
d. ∠ A = 32°   c = 12'   k. ∠ A = 22° 20'   c = 9.5'
e. ∠ A = 19°   c = 14'   l. ∠ A = 26° 30'   c = 11.8'
f. ∠ A = 20°   c = 16'   m. ∠ A = 30° 50'   c = 10.9'
g. ∠ A = 28°   c = 11'   n. ∠ A = 28° 40'   c = 12.6'

3. Find the size of ∠ A in figure 46-10 whose hypotenuse is 8' and base 4'.

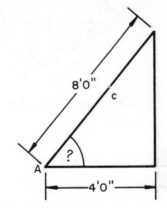

Fig. 46-10

4. Figure 46-11 shows a rectangular floor layout in which two triangles are formed by the diagonal. Solve for the adjacent side (b) with the given measurements.

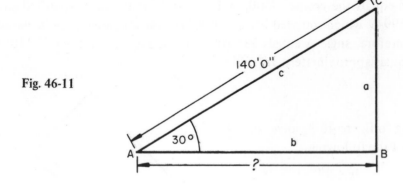

Fig. 46-11

5. If a gable roof has a span of 28' and the length of its rafter is 16', what is the size of the angle formed at the base?

6. A structure has a 20' span, with a 24°30' angle formed by the span and the rafter. What is the length of the rafter?

Unit 47
The Tangent Function

OBJECTIVES

After studying this unit, the student should be able to

- find the value of tangent A in a right triangle.
- find the value of tangent B in a right triangle.
- determine length of sides in a right triangle by using the tangent formula.

DETERMINING TANGENT

The *tangent* of an angle is the ratio of the side opposite that angle to the side adjacent to that angle: opposite side/adjacent side, or a/b. The tangent of angle A may be expressed as tan A.

Illustration. Side a of the right triangle in figure 47-1 measures 6″ and side b measures 8″. Find tangent A.

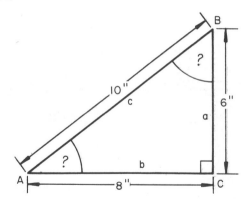

Fig. 47-1 Determining Tangent A and B in a Right Triangle

Solution. Use the formula tan A = a/b.

tan A = 6/8

tan A = .7500

A = approx. 37°

Explanation. Since tan A is the ratio of the side opposite angle A to the side adjacent angle A, the ratio is 6/8. • Substitute 6 for a in the formula and 8 for b. • To find the value of tan A, divide 6 by 8 which equals .7500. • In figure 44-6, the closest tangent value to .7500 is .7490; therefore the approximate measurement of tan A is 37°.

Illustration. The side opposite angle B the right triangle figure 47-1 is 8″, and the side adjacent to angle B is 6″. Find tan B in the right triangle.

Solution. Use the formula, tan B = side opposite angle B/side adjacent angle B

tan B = 8/6

tan B = 1.3333

tan B = approx. 53° 00′

Explanation. • To follow the formula, the side opposite angle B is divided by the side adjacent to angle B. • Therefore, divide 8 by 6 which equals 1.3333. • The figure 1.3333 is located between 1.3270 and 1.3351. • The closest value to 1.3333 is 1.3270. • Therefore, tan B equals approximately 53° 00′.

Notice that the tangent increases as the angle increases.

Illustration. The side adjacent to angle A in figure 47-2A is 12″ and the side opposite the angle is 10″. The side adjacent to angle A in figure 47-2B is 8″ and the side opposite angle A is 4″. Determine if the tan increases as the angle increases.

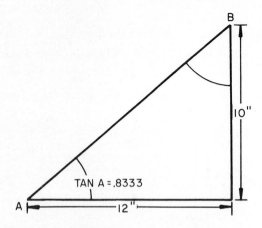

Fig. 47-2A Tangent Increases as Angle Increases.
Tan A (39° 50′) is .8333

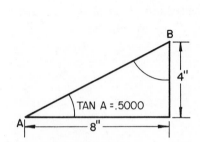

Fig. 47-2B Tan A (26° 30′) is .5000

Solution. Step 1. tan A = a/b

tan A = 10/12

tan A = .8333

Step 2. tan A = a/b

tan A = 4/8

tan A = .5000

Step 3. tan 39° 50′ = approx. .8333

tan 26° 30′ = approx. .5000

Explanation. Since the tangent 39° 50′ equals approximately .8333 and the tangent 26° 30′ equals approximately .5000, it is evident that as the tangent increases, the angle also increases.

DETERMINING ANGLE SIZE

The size of angles in a right triangle can be determined by applying the tangent formula and subtracting that angle from 90°, since there is always a 90° angle in a right triangle. The remaining measurement is the size of the triangle's third angle.

Illustration. Determine the size of the two unknown angles in figure 47-3.

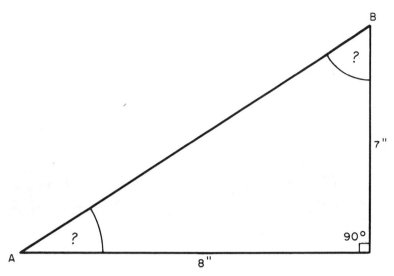

Fig. 47-3 Triangle with Unknown Acute Angles

Solution. Step 1. Apply the formula tan A = a/b

tan A = 7/8

tan A = .8750

tan A = 41° 10′

Step 2. tan B = 90° – 41° 10′

tan B = 48° 50′

angle A = 41° 10′

angle B = 48° 50′

Check. 41° 10′ + 48° 50′ = 90°

90° + 90° = 180°

Explanation. Divide the length of the side opposite ∠ A (7) by the length of the side adjacent ∠ A (8) which equals .8750. Tan .8750 is approximately 41°10′; tan A equals 41°10′. Subtract 41°10′ from 90° which equals 48°50′; therefore, tan B equals 48°50′. To check, add 41°10′ and 48°50′ which equals 90°, the sum of the two acute angles. Add 90° and 90°, equaling 180°. This is the sum of degrees in a right triangle.

DETERMINING SIDE LENGTH

Opposite Side

The length of the opposite side in a right triangle can be determined if ∠ A and the adjacent side are known.

Illustration. A right triangle is formed by the rafters, run, and rise of the roof in figure 47-4. Determine the total rise of the roof with the given measurements.

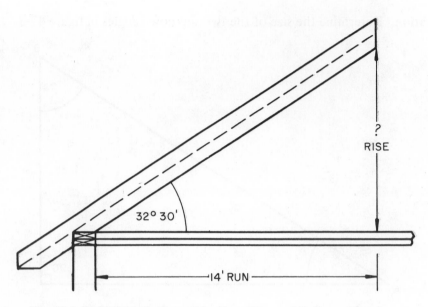

Fig. 47-4 Right Triangle Formed by Run, Rise, and Rafters of Roof

Solution. Step 1. a = rise, b = run

$$a/b = \tan A.$$

$$a = b \times \tan A$$

Step 2. a = 14 x tan 32° 30′

$$\tan 32° 30′ = .63707$$

Step 3. a = 14 x .63707

$$a = 8.91898′$$

Explanation. • To find a, apply the tangent formula: a/b equals tan A. • Multiply b (14′) by the value of tan 32° 30′ (.63707). • The product is 8.91898′. • Rounded off, a equals 9′.

Adjacent Side

The length of the adjacent side can be determined if the opposite side and angle A are known.

Illustration. Find the run of the stair stringer in figure 47-5. There is a 5′ rise and a 30° angle at the base of the stringer.

Fig. 47-5 Staircase Layout

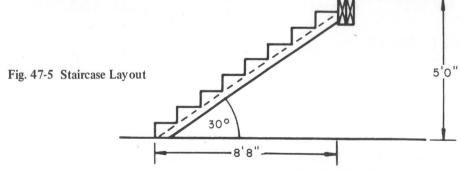

Solution. Use the formula a/b = tan A.

> b = a/tan A
>
> b = 5/tan 30°
>
> b = 5/.57735
>
> b = 8.66025
>
> b = 8.66′
>
> Run = approximately 8′ 8″

Explanation. • The adjacent side equals the opposite side divided by tan A. • Divide 5 by .5774 which, carried out to five decimal places, equals 8.66025 or rounded off, 8′ 8″.

PRACTICAL APPLICATION

A carpenter must determine the size of angles A and B and the length of the rafter. This may be done by applying the tangent trigonometric formula.

Illustration. Find angles A and B to the nearest 10′ and the length of the rafter (hypotenuse), figure 47-6.

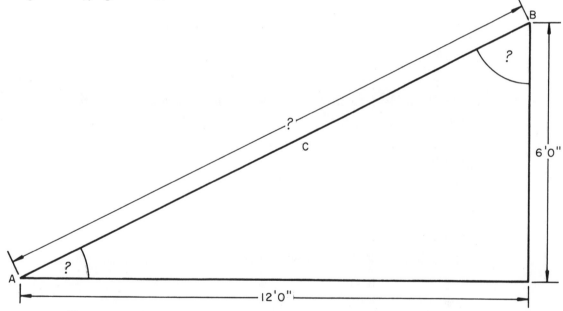

Fig. 47-6 Determine Tan A, Tan B, and Hypotenuse by Tangent Formula.

Solution. Step 1. Use the formula, tan A = a/b

> tan A = 6/12
>
> tan A = .5000
>
> A = approx. 26° 30′ (nearest 10′)

Step 2. B = 90° – A

> B = 90° – 26° 30′
>
> B = approx. 63° 30′ (nearest 10′)

$$\text{Step 3. } c = \sqrt{a^2 + b^2}$$
$$c = \sqrt{6^2 + 12^2}$$
$$c = \sqrt{36 + 144}$$
$$c = \sqrt{180} = 13.4164'$$
$$c = 13.42'$$

Explanation. • Divide 6 by 12 to find the value of tan A (.5000). Tan A equals approx. 26° 30′. • Angle B equals 90° minus 26° 30′ which is approx. 63° 30′. • To find the length of the rafter, find the square root of $6^2 + 12^2$, or 180. The length of the rafter, or c, equals 13.4164′. Rounded off, the answer is 13.42′.

APPLICATION

1. Use figure 44-6 to find the tangent of the following angles

   a. 6°   i. 25°
   b. 13°   j. 30°
   c. 12°   k. 35°
   d. 9°   l. 40°
   e. 7°   m. 45°
   f. 14°   n. 50°
   g. 18°   o. 55°
   h. 20°   p. 60°

2. Use figure 44-6 to find the angles whose tangents have the following values

   a. .5095   g. .6009   m. .2493
   b. 1.9626   h. .6249   n. 4.0108
   c. .5354   i. .1584   o. .2679
   d. 1.8676   j. 6.3138   p. .2867
   e. .5812   k. .1733
   f. 1.7205   l. 5.7694

3. Refer to figure 47-7 and solve for the value of ∠A in each of the following.

   a. a = 4″   b = 6″
   b. a = 6″   b = 8″
   c. a = 7″   b = 9″
   d. a = 9″   b = 7″
   e. a = 5″   b = 4″
   f. a = 3″   b = 6″
   g. a = 3.2″   b = 4.5″
   h. a = 2.5″   b = 3.5″
   i. a = 4.6″   b = 2.9″
   j. a = 5.8″   b = 4.6″
   k. a = 8.8″   b = 10.2″
   l. a = 10.2″   b = 12.4″

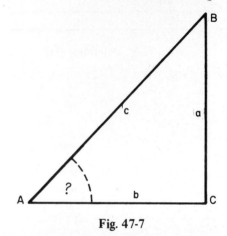

Fig. 47-7

4. Find tan A in the right triangle shown in figure 47-8.

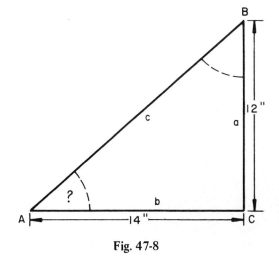

Fig. 47-8

5. Find the angle for tan A of a right triangle whose opposite side is 10' and the adjacent side is 13'.

6. A common rafter has a rise of 6' and a run of 12'. What is the angle measurement of tan A?

7. The rise of a roof is 8' and the angle at the corner of the rafters and the run is 26°. What does the run measure?

8. A corner brace extends up a wall 8' and away from the plate 6'. What is the size of each acute angle in the triangle?

Unit 48
Interpolation

OBJECTIVES

After studying this unit, the student should be able to

- find the value of sines, cosines, and tangents by interpolation.

The carpenter may sometimes find it necessary to determine the angle of a function or the function of an angle which is not listed in figure 44-6, trigonometric functions. When this is the case, interpolation must be used. *Interpolation* involves finding the approximate value of an angle or a function which falls somewhere between two consecutive angles or functions on the chart being used.

When interpolating in the following problems, refer to figure 44-6, trigonometric functions.

INTERPOLATING TO DETERMINE SINES

Illustration. Find the value of sin 29° 18′.

Solution. Step 1. sin 29° 20′ = .48989
sin 29° 10′ = .48735
difference   .00254

Step 2.   29° 18′   29° 20′
   − 29° 10′   − 29° 10′
          8′   10′   Ratio = 8/10

Step 3. 8/10 x .00254 = .02032/10 = .00203

Step 4. .48735 + .00203 = .48938

Explanation. Figure 44-6 does not give the sin of 29°18′, but it does give the sines of 29°20′ and 29°10′. • The sin 29°20′ equals .48998 and the sin 29°10′ equals .48735. Sin 29°18′ is somewhere between 29°20′ and sin 29°10′. • Find the difference between 29°18′ and 29°10′, and 29°20′ and 29°10′, expressed as the ratio 8/10. Find the difference between .48989 and .48735 (.00254). 8/10 of .00254 is .00203; .48735 + .00203 = .48938. Sin 29°18′ = .48938.

INTERPOLATING TO DETERMINE COSINES

Illustration. An *angle brace* fits across angles on the framework of a structure, thereby lending the structure support. Use interpolation to find the cosine expressed on the angle brace in figure 48-1.

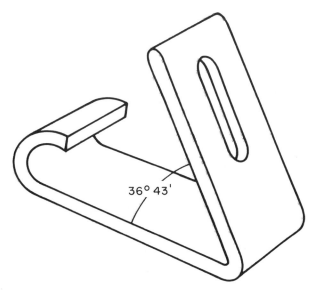

Fig. 48-1 Metal Angle Brace

Solution. Step 1. cos 36°43′ = .80212
            cos 36°50′ = .80038
            difference   .00174

Step 2.   36°43′   36° 50′
           − 36°40′   − 36° 40′
difference   3′   10′   Ratio = 3/10

Step 3. 3/10 x .00174 = .00522/10 = .00052

Step 4. cos 36°43′ = .80212 − .00052   = .80160
            cos 36°43′ = .80160

Explanation. Since 36°43′ is 3/10 of the way between 36°40′ and 36°50′, its cos is about 3/10 of the way between .80212 and .80038. • Multiply 3/10 by .00174 which equals .00052. • Since the cosine of the angle becomes smaller as the angle increases, the difference (.00052) must be subtracted from .80212. • Cos 36° 43′ equals .80160.

INTERPOLATING TO DETERMINE TANGENTS

Illustration. Use interpolation to find tan 18° 12′.

Solution. Step 1. tan 18° 20′ = .33136
            tan 18° 10′ = .32814
            difference   .00322

Step 2.   18° 12′   18° 20′
           − 18° 10′   − 18° 10′
                  2′   10′   Ratio = 2/10

2/10 x .00322 = .00644/ 10 = .00064

Step 3. tan 18° 12′ = .32814 + .00064   = .32878
            tan 18° 12′ = .32878

Explanation. • Tan 18° 12′ is 2/10 of the way between tan 18° 10′ and 18° 20′. • The value of tan 18° 20′ equals .33136 and the value of 18° 10′ equals .32814. • Subtract .32814 from .33136 which equals .00322. • Since the tangent increases as the angle increases, add .32814 and .00064. The sum is .32878. Therefore, tan 18° 12′ = .32878.

APPLICATION

1. Find ∠ A in each of the following functions by interpolation.

   a. sin A = .1693   g. cos A = .9351
   b. sin A = .2093   h. cos A = .9320
   c. sin A = .2714   i. cos A = .9225
   d. sin A = .3076   j. cos A = .9111
   e. sin A = .3434   k. cos A = .9005
   f. sin A = .3814   l. cos A = .9584

2. Find the value of each of the following angles by interpolation.
   a. sin 29° 15′   g. tan 29° 14′
   b. sin 30° 15′   h. tan 51° 12′
   c. sin 6° 45′   i. tan 45° 15′
   d. sin 17° 18′   j. tan 22° 18′
   e. sin 25° 05′   k. tan 72° 45′
   f. sin 64° 43′   l. tan 24° 15′

3. Find ∠ in each of the following functions by interpolation.

   a. tan A = 2.1159   g. sin A = .1406
   b. tan A = 1.6386   h. cos A = .9805
   c. tan A = .3574   i. cos A = .9088
   d. tan A = .1435   j. tan A = .3427
   e. tan A = 1.5956   k. tan A = .3855
   f. tan A = 1.4505   l. tan A = .4057

Glossary of Terms

Acute angle: an angle less than 90° and greater than 0°

Addend: number that is added

Addition: process of combining two or more numbers to form one number

Aggregate: granulated particles used to form a massive substance, such as sand, gravel, and cement

Altitude: the perpendicular height or distance between a point and a lower base

Annex: to add on to

Angle: the figure formed by the intersection of two lines which extend from the same point

Apprentice: one who is learning a trade through an organized program of on-the-job training and school

Area (A): flat surface without thickness

Average: the result obtained by dividing the total of 2 or more quantities by the number of quantities

Batter boards: frames used to lay out corners and partitions of a foundation

Board Foot (bd. ft.): a piece of lumber 1" thick, 1' wide, and 1' long or a piece of material measuring 144 cu. in.

Capacity: content by volume

Circle: a curved figure with all points an equal distance from the center point

Circumference: the perimeter of or linear distance around a circle

Common rafters: a rafter which extends directly from the plate to the ridge

Complex fraction: a fraction having one or more fractions in its numerator and/or denominator

Cone: a solid object whose base is a circle and whose sides taper evenly to a point

Constant: the number in a formula whose value does not change

Cosine (cos): trigonometric function expressing the ratio of the adjacent side to hypotenuse of a right triangle

Corner brace: a piece of material placed diagonally at a corner to reinforce the structure

Cornice: on a roof, trim at the eave of a rafter

Cube solid: object with equal sides, width, height and depth

Cubic foot (cu. ft.): a measurement 1' thick, 1' wide, and 1' long or 1,728 cu. in.

Cubic inch (cu. in.): a measurement 1" thick, 1" wide, and 1" long

Cubic yard (cu. yd.): a measurement 3' thick, 3' wide, and 3' long or 27 cu. ft.

Cylinder: an object bound by two circular ends parallel to each other with a curved surface

Dead load: an inert, permanent load imposed constantly on a structure

Decimal: a proper fraction with a denominator which is a power of 10 and which is placed to the right of a decimal point

Degree (°): a division of a measuring instrument

Denominator: the number in a fraction which is written below the division line

Diameter: the distance equaling twice the radius of a circle

Difference: the result or answer in subtraction

Division: a process of finding the missing factor when one of the two factors are known

Eave: the overhang of a rafter over an exterior wall

Equilateral triangle: a triangle with three angles equal in size

Equivalent: expression of like value

Factor: one number used in a process to obtain another number

Footing: the base of a foundation wall, column, or pier

Formula: a general principle expressed in symbols

Foundation: supportive members of a structure located under the first floor, such as footings and piers

Fraction: in math, a part of a whole

Frustum: formed by cutting off the top of a cone-shaped solid with a plane parallel to the base

Gable roof: a roof formed by common rafters

Gallon (gal.): a unit of liquid measure, equal to 4 quarts

Geometry: a branch of math dealing with the measurement, relationship, and properties of figures, points, lines, and solids

Hip rafter: the rafter which extends from the corner of the building diagonally to the ridge and forms a valley

Horizontal: parallel to the horizon

Hour: measurement of time equal to 60 minutes

Hypotenuse: the side of a right triangle opposite the right angle

Improper fraction: a fraction with a numerator which is the same as the denominator or with a numerator greater than the denominator

Inch (in., ″): a fractional part of one foot

Interest: the sum of money paid to the loaning institution in addition to the money borrowed

Interpolation: the process used to find the approximate value of a function of an angle or an angle of a function between two consecutive angles

Invert: the process by which the numerator of a fraction is placed beneath the division line and the denominator is placed above the line

Joist: a piece of lumber to which floor boards are nailed

Journeyman: a tradesman who is fully qualified in his trade

Lateral area: the outside area of a 3-dimensional object

Layout: marking a given measurement

Least common denominator (L.C.D.): least common multiple of the denominators of two or more fractions

Least common multiple (L.C.M.): the smallest number that will divide each denominator of two or more fractions without a remainder

Linear: measure of length

Live load: moving objects in a building

Metric system: a decimal system of measurement with the meter, liter, and gram as basic measures

Micrometer ("mike"): instrument used to make percise measurements of outside and inside diameters

Minuend: the number from which another number is subtracted

Minute: measure of time equaling 60 seconds

Mixed decimal: a number comprised of a whole number and a decimal

Mixed number: a number comprised of a whole number and a fraction

Multiplicand: a number being multiplied

Multiplication: the process of adding one number to itself a specified number of times

Multiplier: the number in multiplication by which the multiplicand is multiplied

Non-negative number: includes positive numbers and zero

Notation: in math, the expression of numbers in symbols

Numerator: the number in a fraction that is written above the division line

Obtuse angle: an angle greater than 90° and less than 180°

On center (O.C.): spacing of frame members from the center of one member to the center of the member next to it

Ounces (oz.): a weight measure, 16 of which equals 1 lb.

Parallelogram: an object or figure with opposite sides equal and parallel

Parentheses (( )): symbol used in a mathematical problem or equation used to treat an expression as a whole

Percent (%): next to a number, indicates the number of hundredths expressed

Perimeter (P): lengthwise, the distance around an object

Pi (π): 3.1416 or 3 1/7; used to express the ratio of the circumference of a circle to the diameter

Pint (pt.): capacity measure equaling 16 oz.

Pitch: the ratio of the rise of a roof to its span

Polygon: a plane figure bound by 3 or more sides

Pound (lb.): a weight measure equaling 16 oz.

Prime number: a number in which the least common denominator cannot be determined by division

Principal: in a loan, the sum of money borrowed

Product: the result of multiplication

Proper fraction: a fraction in which the numerator is smaller than its denominator

Proportion: in math, the condition in which one ratio is equal to another

Pyramid: an object with its outside walls in the form of triangles which meet in a point at the top

Quart (qt.): in the English system of measurement, a liquid or dry unit equal to 2 pt.

Radius (r): one-half the diameter of a circle; measured from the center of the circle to its perimeter

Ratio: the result obtained by comparing two numbers by division

Rectangle: a parallelogram with four 90° angles

Reflex angle: angle greater than 180° and less than 360°

Remainder: the answer or result in subtraction

Right angle: a 90° angle

Rise: the perpendicular distance between the ridge of a roof and the plate

Riser: the vertical height on a stair stringer between two consecutive treads

Run: distance equal to one-half the width of a building

Second: a measure of time, 60 of which equals 1 minute; also, an angular measure, 60 of which equals 1 minute

Sine (sin): trigonometric function expressing the ratio of the side opposite the right angle to the hypotenuse of a right triangle

Span: the total width of a building

Sphere: a globular object, such as a ball

Square (sq.): a square measurement equal to 100 sq. ft.

Square foot (sq. ft.): a square measurement equal to 144 sq. in.

Square inch (sq. in.): a measurement of 1″ on four sides

Square root: the number when multiplied by itself equals the given number

Square yard (sq. yd.): a measurement of 3′ by 3′ or 9 sq. ft.

Stringer: a horizontal structural member of a stair

Subtraction: the process of taking one number from another number to obtain a resulting number

Subtrahend: in subtraction, the number subtracted from another number

Sum: the result or answer in addition

Tangent (tan): trigonometric function expressing the ratio of the opposite side of a right triangle to the adjacent side

Terms: in math, the numerator and denominator of a fraction

Ton, long (l.t.): measure of weight (avoirdupois) equaling 2240 lb.

Ton, short (s.t.): measure of weight (avoirdupois) equaling 2000 lb.

Transpose: in math, to move a term from one side of an equation to the other

Trapezoid: a figure with four straight sides, only two of which are parallel

Triangle: a geometric figure formed by straight lines and with three angles

Trigometry: a branch of mathematics in which unknown angles and sides of triangles are determined

Trignometric functions: expressions showing the ratio of two sides of a triangle

Unit: a standard by which length, area, and volume are measured

Vertex: the point where the lines of an angle meet

Vertical: perpendicular to the plane of the horizon

Volume (vol.): one type of measurement; also known as capacity

Weight (wt.); the straight downward pull on an object by gravity

Yard (yd.): linear measure equaling 3′

Appendix

STANDARD TABLES OF ENGLISH MEASURE

Linear Measure		
12 inches (in.)	=	1 foot (ft).
3 ft.	=	1 yard (yd.)
16 1/2 ft.	=	1 rod (rd.)
5 1/2 yd.	=	1 rd.
320 rd.	=	1 mile
1760 yd.	=	1 mile
5280 ft.	=	1 mile

Surface Measure		
144 sq. in.	=	1 sq. ft.
9 sq. ft.	=	1 sq. yd.
30 1/4 sq. yd.	=	1 sq. rd.
160 sq. rd.	=	1 acre
640 acres	=	1 sq. mile
43,560 sq. ft.	=	1 acre

Cubic Measure		
1728 cu. in.	=	1 cu. ft.
27 cu. ft.	=	1 cu. yd.
128 cu. ft.	=	1 cord

Angular (Circular) Measure		
60 sec. ('')	=	1 min. (')
60'	=	1 degree (°)
90°	=	1 quadrant
360°	=	1 circle

Time Measure		
60 seconds (sec.)	=	1 minute (min.)
60 min.	=	1 hour (hr.)
24 hr.	=	1 day
7 days	=	1 week
52 weeks	=	1 year
365 days	=	1 year
10 years	=	1 decade

Liquid Measure		
4 gills	=	1 pint (pt.)
2 pt.	=	1 quart (qt.)
4 qt.	=	1 gallon (gal.)
231 cu. in.	=	1 gal.
31.5 gal.	=	1 barrel (bbl.)
42 gal.	=	1 bbl. of oil
8 1/2 lb.	=	1 gal. water
7 1/2 gal.	=	1 cu. ft.

Weights of Materials		
0.096 lb.	=	1 cu. in. aluminum
0.260 lb.	=	1 cu. in. cast iron
0.283 lb.	=	1 cu. in. mild steel
0.321 lb.	=	1 cu. in. copper
0.41 lb.	=	1 cu. in. lead
112 lb.	=	1 cu. ft. Dowmetal
167 lb.	=	1 cu. ft. aluminum
464 lb.	=	1 cu. ft. cast iron
490 lb.	=	1 cu. ft. mild steel
555.6 lb.	=	1 cu. ft. copper
710 lb.	=	1 cu. ft. lead

Avoirdupois Weight		
16 ounces (oz.)	=	1 pound (lb.)
100 lb.	=	1 hundredweight (cwt.)
20 cwt.	=	1 ton
2000 lb.	=	1 ton
8 1/2 lb.	=	1 gal. of water
62.4 lb.	=	1 cu. ft. of water
112 lb.	=	1 long cwt.
2240 lb.	=	1 long ton

Dry Measure		
2 cups	=	1 pt.
2 pt.	=	1 qt.
4 qt.	=	1 gal.
8 qt.	=	1 peck (pk.)
4 pk.	=	1 bushel (bu.)

Miscellaneous		
12 units	=	1 dozen (doz.)
12 doz.	=	1 gross
144 units	=	1 gross
24 sheets	=	1 quire
20 quires	=	1 ream
20 units	=	1 score
6 ft.	=	1 fathom

CONVERSION OF ENGLISH AND METRIC MEASURES

Linear Measure

Unit	Inches to milli-meters	Milli-meters to inches	Feet to meters	Meters to feet	Yards to meters	Meters to yards	Miles to kilo-meters	Kilo-meters to miles
1	25.40	0.03937	0.3048	3.281	0.9144	1.094	1.609	0.6214
2	50.80	0.07874	0.6096	6.562	1.829	2.187	3.219	1.243
3	76.20	0.1181	0.9144	9.842	2.743	3.281	4.828	1.864
4	101.60	0.1575	1.219	13.12	3.658	4.374	6.437	2.485
5	127.00	0.1968	1.524	16.40	4.572	5.468	8.047	3.107
6	152.40	0.2362	1.829	19.68	5.486	6.562	9.656	3.728
7	177.80	0.2756	2.134	22.97	6.401	7.655	11.27	4.350
8	203.20	0.3150	2.438	26.25	7.315	8.749	12.87	4.971
9	228.60	0.3543	2.743	29.53	8.230	9.842	14.48	5.592

Example   1 in. = 2540 mm.,   1 m. = 3.281 ft.,   1 Km. = 0.6214 mi.

Surface Measure

Unit	Square inches to square centi-meters	Square centi-meters to square inches	Square feet to square meters	Square meters to square feet	Square yards to square meters	Square meters to square yards	Acres to hec-tares	Hec-tares to acres	Square miles to square kilo-meters	Square kilo-meters to square miles
1	6.452	0.1550	0.0929	10.76	0.8361	1.196	0.4047	2.471	2.59	0.3861
2	12.90	0.31	0.1859	21.53	1.672	2.392	0.8094	4.942	5.18	0.7722
3	19.356	0.465	0.2787	32.29	2.508	3.588	1.214	7.413	7.77	1.158
4	25.81	0.62	0.3716	43.06	3.345	4.784	1.619	9.884	10.36	1.544
5	32.26	0.775	0.4645	53.82	4.181	5.98	2.023	12.355	12.95	1.931
6	38.71	0.93	0.5574	64.58	5.017	7.176	2.428	14.826	15.54	2.317
7	45.16	1.085	0.6503	75.35	5.853	8.372	2.833	17.297	18.13	2.703
8	51.61	1.24	0.7432	86.11	6.689	9.568	3.237	19.768	20.72	3.089
9	58.08	1.395	0.8361	96.87	7.525	10.764	3.642	22.239	23.31	3.475

Example   1 sq. in. = 6.452 sq. cm.,   1 sq. m. = 1.196 sq. yds., 1 sq. mi. = 2.59 sq. Km.

Cubic Measure

Unit	Cubic inches to cubic centi-meters	Cubic centi-meters to cubic inches	Cubic feet to cubic meters	Cubic meters to cubic feet	Cubic yards to cubic meters	Cubic meters to cubic yards	Gallons to cubic feet	Cubic feet to gallons
1	16.39	0.06102	0.02832	35.31	0.7646	1.308	0.1337	7.481
2	32.77	0.1220	0.05663	70.63	1.529	2.616	0.2674	14.96
3	49.16	0.1831	0.08495	105.9	2.294	3.924	0.4010	22.44
4	65.55	0.2441	0.1133	141.3	3.058	5.232	0.5347	29.92
5	81.94	0.3051	0.1416	176.6	3.823	6.540	0.6684	37.40
6	98.32	0.3661	0.1699	211.9	4.587	7.848	0.8021	44.88
7	114.7	0.4272	0.1982	247.2	5.352	9.156	0.9358	52.36
8	131.1	0.4882	0.2265	282.5	6.116	10.46	1.069	59.84
9	147.5	0.5492	0.2549	371.8	6.881	11.77	1.203	67.32

Example   1 cu. cm. = 0.06102 cu. in.,   1 gal. = 0.1337 cu. ft.

Volume or Capacity Measure

Unit	Liquid ounces to cubic centi-meters	Cubic centi-meters to liquid ounces	Pints to liters	Liters to pints	Quarts to liters	Liters to quarts	Gallons to liters	Liters to gallons	Bushels to hecto-liters	Hecto-liters to bushels
1	29.57	0.03381	0.4732	2.113	0.9463	1.057	3.785	0.2642	0.3524	2.838
2	59.15	0.06763	0.9463	4.227	1.893	2.113	7.571	0.5284	0.7048	5.676
3	88.72	0.1014	1.420	6.340	2.839	3.785	11.36	0.7925	1.057	8.513
4	118.3	0.1353	1.893	8.454	3.170	4.227	15.14	1.057	1.410	11.35
5	147.9	0.1691	2.366	10.57	4.732	5.284	18.93	1.321	1.762	14.19
6	177.4	0.2029	2.839	12.68	5.678	6.340	22.71	1.585	2.114	17.03
7	207.0	0.2367	3.312	14.79	6.624	7.397	26.50	1.849	2.467	19.86
8	236.6	0.2705	3.785	16.91	7.571	8.454	30.28	2.113	2.819	22.70
9	266.2	0.3043	4.259	19.02	8.517	9.510	34.07	2.378	3.171	25.54

Example   1 l. = 2.113 pts.,   1 gal. = 3.785 l.

POWERS AND ROOTS OF NUMBERS (1 through 100)

Num-ber	Powers		Roots		Num-ber	Powers		Roots	
	Square	Cube	Square	Cube		Square	Cube	Square	Cube
1	1	1	1.000	1.000	51	2,601	132,651	7.141	3.708
2	4	8	1.414	1.260	52	2,704	140,608	7.211	3.733
3	9	27	1.732	1.442	53	2,809	148,877	7.280	3.756
4	16	64	2.000	1.587	54	2,916	157,464	7.348	3.780
5	25	125	2.236	1.710	55	3,025	166,375	7.416	3.803
6	36	216	2.449	1.817	56	3,136	175,616	7.483	3.826
7	49	343	2.646	1.913	57	3,249	185,193	7.550	3.849
8	64	512	2.828	2.000	58	3,364	195,112	7.616	3.871
9	81	729	3.000	2.080	59	3,481	205,379	7.681	3.893
10	100	1,000	3.162	2.154	60	3,600	216,000	7.746	3.915
11	121	1,331	3.317	2.224	61	3,721	226,981	7.810	3.936
12	144	1,728	3.464	2.289	62	3,844	238,328	7.874	3.958
13	169	2,197	3.606	2.351	63	3,969	250,047	7.937	3.979
14	196	2,744	3.742	2.410	64	4,096	262,144	8.000	4.000
15	225	3,375	3.873	2.466	65	4,225	274,625	8.062	4.021
16	256	4,096	4.000	2.520	66	4,356	287,496	8.124	4.041
17	289	4,913	4.123	2.571	67	4,489	300,763	8.185	4.062
18	324	5,832	4.243	2.621	68	4,624	314,432	8.246	4.082
19	361	6,859	4.359	2.668	69	4,761	328,509	8.307	4.102
20	400	8,000	4.472	2.714	70	4,900	343,000	8.367	4.121
21	441	9,261	4.583	2.759	71	5,041	357,911	8.426	4.141
22	484	10,648	4.690	2.802	72	5,184	373,248	8.485	4.160
23	529	12,167	4.796	2.844	73	5,329	389,017	8.544	4.179
24	576	13,824	4.899	2.884	74	5,476	405,224	8.602	4.198
25	625	15,625	5.000	2.924	75	5,625	421,875	8.660	4.217
26	676	17,576	5.099	2.962	76	5,776	438,976	8.718	4.236
27	729	19,683	5.196	3.000	77	5,929	456,533	8.775	4.254
28	784	21,952	5.292	3.037	78	6,084	474,552	8.832	4.273
29	841	24,389	5.385	3.072	79	6,241	493,039	8.888	4.291
30	900	27,000	5.477	3.107	80	6,400	512,000	8.944	4.309
31	961	29,791	5.568	3.141	81	6,561	531,441	9.000	4.327
32	1,024	32,798	5.657	3.175	82	6,724	551,368	9.055	4.344
33	1,089	35,937	5.745	3.208	83	6,889	571,787	9.110	4.362
34	1,156	39,304	5.831	3.240	84	7,056	592,704	9.165	4.380
35	1,225	42,875	5.916	3.271	85	7,225	614,125	9.220	4.397
36	1,296	46,656	6.000	3.302	86	7,396	636,056	9.274	4.414
37	1,369	50,653	6.083	3.332	87	7,569	658,503	9.327	4.481
38	1,444	54,872	6.164	3.362	88	7,744	681,472	9.381	4.448
39	1,521	59,319	6.245	3.391	89	7,921	704,969	9.434	4.465
40	1,600	64,000	6.325	3.420	90	8,100	729,000	9.487	4.481
41	1,681	68,921	6.403	3.448	91	8,281	753,571	9.539	4.498
42	1,764	74,088	6.481	3.476	92	8,464	778,688	9.592	4.514
43	1,849	79,507	6.557	3.503	93	8,649	804,357	9.644	4.531
44	1,936	85,184	6.633	3.530	94	8,836	830,584	9.695	4.547
45	2,025	91,125	6.708	3.557	95	9,025	857,375	9.747	4.563
46	2,116	97,336	6.782	3.583	96	9,216	884,736	9.798	4.579
47	2,209	103,823	6.856	3.609	97	9,409	912,673	9.849	4.595
48	2,304	110,592	6.928	3.634	98	9,604	941,192	9.900	4.610
49	2,401	117,649	7.000	3.659	99	9,801	970,299	9.950	4.626
50	2,500	125,000	7.071	3.684	100	10,000	1,000,000	10.000	4.642

ARCHITECTURAL SYMBOLS

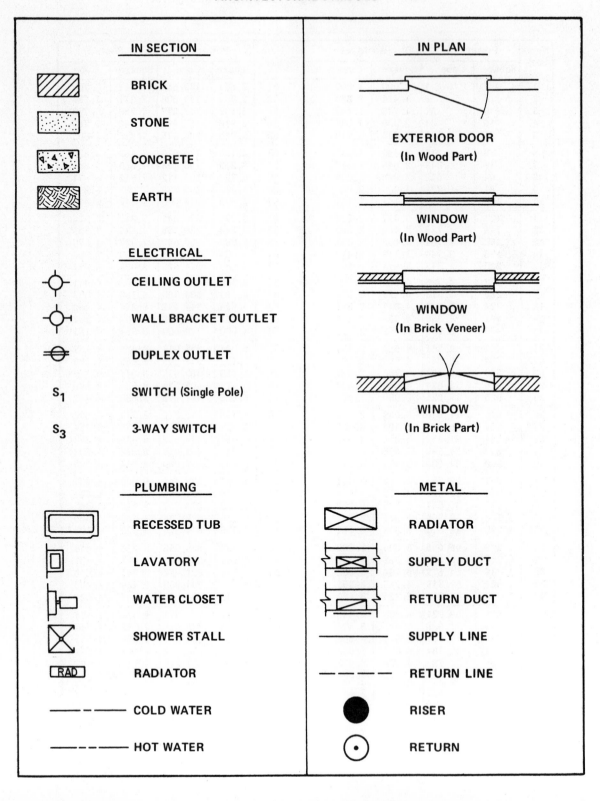

IN SECTION

BRICK

STONE

CONCRETE

EARTH

ELECTRICAL

CEILING OUTLET

WALL BRACKET OUTLET

DUPLEX OUTLET

S_1 SWITCH (Single Pole)

S_3 3-WAY SWITCH

PLUMBING

RECESSED TUB

LAVATORY

WATER CLOSET

SHOWER STALL

RADIATOR

COLD WATER

HOT WATER

IN PLAN

EXTERIOR DOOR
(In Wood Part)

WINDOW
(In Wood Part)

WINDOW
(In Brick Veneer)

WINDOW
(In Brick Part)

METAL

RADIATOR

SUPPLY DUCT

RETURN DUCT

SUPPLY LINE

RETURN LINE

RISER

RETURN

ARCHITECTURAL SYMBOLS

PLUMBING SYMBOLS

ROLL RIM
TUB

ANGLE
TUB

SHOWER
STALL

WATER
CLOSET

BIDET

URINAL
STALL TYPE

LAVATORY

KITCHEN SINK
R & L DRAIN BOARD

COMBINATION SINK
AND DISHWASHER

COMBINATION SINK
AND LAUNDRY TRAY

S   T

HOT WATER
TANK

H W T

WATER METER

M

HOSE BIBB
OR FAUCET

DRAIN

D

GAS RANGE

R

DRYER

D

WASHING
MACHINE

WM

WALL–TYPE
DRINKING FOUNTAIN

DRY
WELL

DW

WATER HEATER

WH

HEATING AND VENTILATING

RADIATOR
EXPOSED

RAD

UNIT HEATER

UNIT VENTILATOR

TRAP
THERMOSTATIC

TRAP – FLOAT AND
THERMOSTAT

TRAP
BOILER RETURN

VALVE AIR LINE

VALVE
DIAPHRAGM

VALVE
STRAINER

THERMOSTAT

T

Acknowledgments

Contributions by Delmar Staff

    Publication Director — Alan N. Knofla

    Source Editor — Mary R. Grauerholz

    Editor, Instructor's Guide — John Rosencrans

    Copy Editor — Sharon Patnode

    Photographer — Richard T. Kreh

    Director of Manufacturing and Production — Fred Sharer

    Illustrators — Michael Kokernak, George Dowse, Anthony Canabush

    Production Specialists — Lee St.Onge, Sharon Lynch, Patti Barosi, Betty Michelfelder, Jean LeMorta

Contributions of Content and Illustrations

    Dr. R.W.E. Jones
    Dr. E.L. Cole
    Mrs. Judia Hobdy
    Mrs. Rebecca D. Lewis
    Mr. George Glenn
    Mr. S.U. Walton
    Mr. Emmett Johnson
    Stanley Tool Company
    L.S. Starret Company
    U.S. Office of Education
    Students of Industrial Technology, Grambling State University
    MANPOWER Magazine
    National Bureau of Standards

Index

Answers to Odd Numbered Questions

SECTION 1 WHOLE NUMBERS

Unit 1 Reading and Writing Whole Numbers

APPLICATION, Pages 4-6

1. a. +, plus, and   b. −, minus, less   c. x, ·, ( ), times   d. ÷, ⌐ ,)⎯

3.
a. one	f. six	k. eleven	p. sixteen
b. two	g. seven	l. twelve	q. seventeen
c. three	h. eight	m. thirteen	r. eighteen
d. four	i. nine	n. fourteen	s. nineteen
e. five	j. ten	o. fifteen	t. twenty

5.
a. 25	e. 99	h. 88	k. 67	n. 98	q. 47	t. 74
b. 16	f. 85	i. 75	l. 64	o. 30	r. 58	u. 90
c. 19	g. 72	j. 66	m. 86	p. 33	s. 97	v. 38
d. 23						

7. a. 57,057,010   b. 7,800   c. 13,000,215   d. 3,050   e. 100,989

9.
a. 5,809 sq. in.	f. 950,670 sq. yd.	k. 9,100,000 cu. yd.
b. 2,579 sq. in.	g. 867,059 sq. yd.	l. 8,105,600 cu. yd.
c. 9,349 sq. ft.	h. 765,805 sq. yd.	m. 50,687,958 cu. ft.
d. 9,506 sq. ft.	i. 658,999 cu. in.	n. 411,224,060,121 lbs.
e. 10,700 sq. ft.	j. 100,000 cu. in.	o. 521,232,760,326 lbs.

Unit 2 Addition of Whole Numbers

APPLICATION, Pages 9-10

1.
a. 6	d. 21	g. 17	j. 21	m. 18	p. 15
b. 9	e. 20	h. 22	k. 18	n. 16	q. 15
c. 15	f. 19	i. 22	l. 12	o. 11	r. 25

3.
a. 45	c. 56	e. 22	g. 32
b. 53	d. 46	f. 32	h. 35

5. 1881   22,706   14,860   26,190

7. $20,650

Unit 3 Subtraction of Whole Numbers

APPLICATION, Pages 15-16

1. 14'

3.
0	2	4	4	6
0	1	6	6	3
4	4	1	8	5

5.
61	9	131	289	734	1185
592	16	106	182	111	901

7. 1,169 bricks

9. 40"

11.
28	41	53	2,200	594	240,136
14	32	38	1,089	635,790	140,458

Unit 4 Multiplication of Whole Numbers

APPLICATION, Pages 20-21

1.
a. 405	g. 783	l. 210	q. 344	v. 229,360
b. 332	h. 328	m. 855	r. 9,308	w. 1,672
c. 348	i. 364	n. 738	s. 1,876	x. 238,280
d. 228	j. 490	o. 124	t. 246,392	y. 114,452
e. 644	k. 447	p. 395	u. 102,600	z. 43,343
f. 170				

3. 1,102 sq. ft.

5. $115.00

7. $160.00

Unit 5 Division of Whole Numbers

APPLICATION, Pages 25-27

1.
a. 4	b. 6	c. 11 2/3
32 dividend	72 dividend	105 dividend
8 divisor	12 divisor	9 divisor
4 quotient	6 quotient	11 quotient
no remainder	no remainder	2/3 remainder

3. 2 9/14"

5.
a. 1 R 66	d. 4 R 9	g. 13	j. 1 R 147
b. 2 R 60	e. 5	h. 33	k. 20 R 78
c. 6 R 41	f. 19	i. 108 R 9	l. 85 R 360

7.
a. 8	b. 42	c. 396	d. 157.15

9. 12"

11. 12

13. Place rule diagonally across the board to 12" and mark at 4 and 8.

SECTION 2 FRACTIONS

Unit 6 Addition of Fractions

APPLICATION, Pages 33-35

1.
a. 1/4	f. 7/8	k. 1/3	p. 6/24	t. 6/36
b. 2/4	g. 1/5	l. 2/16	q. 4/10	u. 9/38
c. 1/3	h. 8/16	m. 5/20	r. 15/22	v. 10/24
d. 2/3	i. 1/2	n. 6/26	s. 7/9	w. 15/25
e. 5/6	j. 3/5	o. 6/21		

3. 4 1/2"

5.
a. 1	b. 1/2	c. 1/4	d. 1/3	e. 3/4	f. 3/8

7. 6 9/16"

9. 2 1/2"

11.
a. 25/30	e. 60/120	i. 18/48	m. 6/12	q. 2/16
b. 9/12	f. 10/16	j. 44/64	n. 6/12	r. 30/64
c. 6/12	g. 27/72	k. 9/27	o. 38/64	s. 15/30
d. 30/32	h. 4/12	l. 16/64	p. 20/32	t. 70/100

Unit 7 Subtraction of Fractions

APPLICATION, Pages 38-41

1. a. 17/24   d. 9/64   g. 15/32   j. 13/64   m. 3/16
    b. 27/64   e. 3/16   h. 5/16   k. 1/35   n. 11/40
    c. 3/10   f. 8/35   i. 19/64   l. 5/16   o. 56/143

3. 3/8"

5. a. 1/10   f. 13/40   j. 23/80   n. 7/20   r. 1/20
    b. 13/24   g. 7/20   k. 4/15   o. 5/24   s. 17/42
    c. 1/18   h. 37/60   l. 5/36   p. 1/8   t. 1/15
    d. 3/10   i. 21/40   m. 3/6   q. 1/40   u. 7/40
    e. 1/20

7. 1/32"

9. 3/4"

Unit 8 Multiplication of Fractions

APPLICATION, Page 48

1. a. 3/8   b. 15/64   c. 4/15   d. 9/98   e. 957/2048   f. 63/1710

3. 5 1/3

5. a. 12 bd. ft.   b. 200 bd. ft.   c. 10 bd. ft.   d. 30 bd. ft.   e. 24 bd. ft.   f. 10 bd. ft.

7. 6 ft.

9. 6 ft. rise

Unit 9 Division of Fractions

APPLICATION, Pages 50-53

1. a. 1 1/4   e. 8 2/3   i. 17/23   m. 16/21   q. 3 3/5
    b. 12   f. 4/5   j. 5/11   n. 1 1/17 r. 21/64
    c. 1 11/21 g. 1/6   k. 1 1/3   o. 2 3/5   s. 12/17
    d. 1 1/2   h. 1/4   l. 2 2/5   p. 6/13   t. 1/2

3. a. 24   f. 66   k. 56   o. 64   s. 56 2/3
    b. 40   g. 18   l. 20 1/4 p. 500   t. 130 2/3
    c. 57   h. 182   m. 36   q. 32   u. 170 2/3
    d. 8   i. 16   n. 60   r. 6   v. 16 5/7
    e. 19 4/5 j. 20

5. 4 shelves

7. a. 32   e. 20   i. 64   m. 27   p. 10 2/3
    b. 16 1/3 f. 36 2/3 j. 10   n. 10 1/2 q. 21 1/3
    c. 16 2/3 g. 22 2/5 k. 13 1/2 o. 45   r. 56 8/9
    d. 27 3/7 h. 48   l. 32

9. 3

11. 48 pcs.

13. 48 ft.

15. 49 studs

Unit 10 Addition of Mixed Numbers

APPLICATION, Pages 56-59

1.
a. 1 1/3	g. 1 3/4	m. 2 5/14	r. 4 12/13	w. 1 3/5	ab. 2
b. 2 1/2	h. 2	n. 1 3/5	s. 16	x. 7 1/2	ac. 4 5/11
c. 2	i. 3 1/3	o. 3 1/5	t. 1 1/5	y. 1 1/2	ad. 5
d. 2	j. 1 1/16	p. 4	u. 2	z. 1 11/28	ae. 9
e. 2 2/3	k. 2 7/8	q. 4 4/7	v. 3	aa. 2	af. 10
f. 1 2/3	l. 5				

3.
a. 1 5/16	c. 1 21/40	e. 49/64	g. 1 2631/3120
b. 1 5/16	d. 1 23/36	f. 1 127/288	h. 1003/1755

5.
a. 16 1/2	c. 7 9/10	e. 10 1/4	g. 7 8/9
b. 10 1/15	d. 27 19/24	f. 23 19/24	h. 8 5/6

7. width = 32 11/16   length = 56 1/8

9. a. 13 5/8′ cast 4 in.   b. 12 15/8′ cast 2 in.

11. 261 3/16 lbs.

13. 7 5/32 in.

15. 13 17/20 days

Unit 11 Subtraction of Mixed Numbers

APPLICATION, Pages 63-64

1.	14 1/4″	13. 1 5/12 hr.
3.	14 3/8″	15. 2 11/24 hr.
5.	444 19/20 bd. ft.	17. 2 7/8″
7.	1 43/48	19. 2 1/4′
9.	1/2	21. 55 3/16 hr.
11.	3 1/4″	

Unit 12 Multiplication of Mixed Numbers

APPLICATION, Pages 69-70

1.
a. 25 1/2	e. 28 1/2	i. 649 1/7	m. 47	q. 39 3/5
b. 40 1/2	f. 134 1/16	j. 20	n. 27	r. 16 2/3
c. 16 13/32	g. 17	k. 71 3/4	o. 14	
d. 52	h. 35 5/8	l. 4 7/8	p. 135 3/16	

3. 6 3/4″

5. 1,798 bricks

Unit 13 Division of Mixed Numbers

APPLICATION, Pages 73-76

1.
a. 2 2/3	e. 3 5/9	i. 2 5/8	m. 2 1/12	q. 6 3/4	u. 2 2/11
b. 15 5/11	f. 12	j. 6 2/37	n. 3 11/23	r. 2 74/75	v. 2 10/11
c. 6 6/17	g. 16 8/9	k. 18 3/19	o. 11 5/9	s. 21 1/195	w. 21 3/5
d. 3 13/21	h. 3 1/13	l. 15 9/17	p. 2 17/19	t. 9 3/5	x. 42 50/203

3. 36 pieces

5. 7 1/3 hr.

7.
a. 3 30/31	f. 13/21	j. 1 1/24	n. 37/57	r. 1 23/36
b. 1 4/5	g. 3 13/37	k. 1	o. 9 73/83	s. 1
c. 26/35	h. 2 7/39	l. 6 14/17	p. 1 11/38	t. 1 11/117
d. 3 7/9	i. 45/52	m. 3 1/2	q. 81/116	u. 1 37/68
e. 1 3/25				

9. 20 5/6 or 21

11. 5 26/35

13. 40 sq. ft.

15. 13'

17. 240

Unit 14 Conversion of Decimal Fractions to Common Fractions

APPLICATION, Page 82

1.
One tenth	Two hundredths	Three thousandths	Thirty-seven hundredths
Two tenths	Three hundredths	Four thousandths	Eighty-five hundredths
Six tenths	Seven hundredths	Eight thousandths	
Seven tenths	Eight hundredths	Nine thousandths	

3.
a. 1/4	d. 19/100	g. 3/40	j. 9/2000	m. 473/5000
b. 1/2	e. 9/40	h. 3/4	k. 7/80	n. 3/8
c. 9/20	f. 69/200	i. 9/10	l. 3/200	o. 223/400

5. a. 56/64   b. 7/8   c. 1/8

Unit 15 Conversion of Fractions to Decimal Fractions

APPLICATION, Pages 87-89

1.
a. 1.50	i. 2.40625	p. 6.078125	w. 5.0140625	ad. 7.021875
b. 5.3125	j. 5.265625	q. 2.0625	x. 11.0078125	ae. 11.0109375
c. 11.21875	k. 7.1875	r. 7.03125	y. 8.0500	af. 5.0900
d. 5.1250	l. 4.09375	s. 5.015625	z. 10.0250	ag. 3.0375
e. 2.46875	m. 4.3750	t. 5.00859	aa. 6.0125	ah. 2.01875
f. 2.28125	n. 10.2500	u. 3.05625	ab. 9.007032	ai. 3.01015625
g. 7.203125	o. 5.171875	v. 9.028125	ac. 7.04375	aj. 2.1100
h. 5.109375				

3. .375"

5. 1.375"

7. 31/64

9. .8125 ft.

11. 2.5104 ft.

13. See appendix decimal equivalents

15. .750 lbs.

Unit 16 Addition of Decimals

APPLICATION, Pages 93-95

1. a. 6.70   e. 7.129   i. 9.777   m. 10.172   q. 547.607
    b. 1.226   f. 7.921   j. 8.7782   n. 4.4656   r. 1.4312
    c. 8.41   g. 2.27   k. 8.160   o. 2102.10056 s. 39.182
    d. 2.2841   h. 9.049   l. 61.2284   p. 9.912

3. .156"

5. a. 2.1875   b. 1.6813   c. 1.1295   d. 2.4573

7. 92.925 lbs.

9. 53.485 lbs.

11. a. .412"   b. .519"   c. .708"   d. .301"   e. .818"   f. .613"

13. 25.783 thousands

Unit 17 Subtraction of Decimals

APPLICATION, Pages 98-101

1. a. .5   h. .5   o. .57   v. .66   ab. .29   ah. .151
    b. .1   i. .6   p. .37   w. .22   ac. .06   ai. .692
    c. .4   j. .3   q. .19   x. .15   ad. .17   aj. .219
    d. .5   k. .77   r. .87   y. .40   ae. .346 ak. .180
    e. .3   l. .55   s. .86   z. .11   af. .199 al. .177
    f. .1   m. .67   t. .75   aa. .09   ag. .103 am. .9873
    g. .1   n. .48   u. .09

3. 26.80   15. 4.334'
5. 49.63 ft.   17. .775
7. .1245 cu. yd.   19. 3.784'
9. yes, .011" too large   21. 5.622
11. .980"   23. 4.500"
13. 167.249'

Unit 18 Multiplication of Decimals

APPLICATION, Page 106

1. a. 1.35   d. .4630   g. .0020   j. .0768   m. 306.25
    b. $0.18   e. .4320   h. .30   k. 1.9551   n. 3.231250
    c. $6.07   f. $3.22   i. 3.575   l. .1238297 o. 1.5165625

3. 16.0 sq. ft.

5. $22.5675 or $22.57

7. a. $0.40   b. $0.7719 or $0.77   c. $0.714 or $0.71   d. $5.46   e. $1.875 or $1.88

9. 16.82 ft. common rafter   21.92 ft. hip rafter

11. 17.46 in. per foot run

13. 20.80 in. per foot run

15. $10.5392 or $10.54

Unit 19 Division of Decimals

APPLICATION, Pages 109-111

1. a. 1.0948   d. 133.33   g. 1.27   j. 4960.
   b. 354.55   e. 3.36   h. .31   k. .571
   c. .583   f. 31.68   i. .048

3. 200 bolts   13.   20.39 ft.
5. $56.58   15.   2 pieces
7. 7.33 lbs.   17.   17 pieces
9. 17 days   19.   16 studs
11. 16.970 ft.

SECTION 3 RATIO, PROPORTION, AND PERCENT

Unit 20 Ratio, Proportion and Averages

APPLICATION, Pages 118-119

1. 3/5

3. 3:1

5. a. 1 1/3   c. 200   e. 30   g. 16 4/5   i. 4
   b. 7   d. 2/3   f. 3 3/5   h. 100   j. 175

7. a. 2 9/24   d. 2 117/192   g. 7 11/54
   b. 3 7/48   e. 2 89/96   h. 5 171/192
   c. 2 67/96   f. 2 83/135   i. 3 7/18

9. 9' 4"

11. $11,812

13. 8 44/96

Unit 21 Determining Percents and Quantity

APPLICATION, Pages 125-126

1. a. .04   g. .19   m. 1.057   s. .06 1/4 or .0625
   b. .07   h. 1.25   n. 3.50   t. .064 1/3 or .06433
   c. .74   i. 1.40   o. .50   u. .00 2/7
   d. .44   j. .1533   p. .00 1/3 or .0033   v. .00 1/13
   e. .07   k. 3.15   q. .006 1/6 or .0066   w. .04 2/9 or .0422
   f. .09   l. .14   r. .22   x. .99

3. a. .5 = 50%   g. .875 = 87.5%   m. .15625 = 15.625%
   b. .8 = 80%   h. .6875 = 68.75%   n. .833 1/3 = 83.3 1/3%
   c. .5625 = 56.25%   i. 3.883 1/3 = 383.3 1/3% o. 1.25 = 125%
   d. 5.375 = 537.5%   j. 3.714 2/7 = 371.4 2/7% p. .5 = 50%
   e. .71 4/7 = 71 4/7%   k. .375 = 37.5%   q. 3.375 = 337.5%
   f. .125 = 12.5%   l. .777 7/9 = 77.7 7/9%   r. 1.5 = 150%

5. 21.46'
7. 1.61 inches
9. 5040 lbs.
11. 18 sheets
13. 25%

Unit 22 Percent and Money Management

APPLICATION, Pages 131-132

1.	14,000 bd. ft.	9.	6 months
3.	$2,400	11.	$291
5.	$61.92	13.	$190
7.	40%		

SECTION 4 THE ENGLISH SYSTEM OF MEASUREMENT

Unit 23 Using the English System: Linear, Volume, and Weight Measurements

APPLICATION, Pages 142-143

1.
a. 48″	d. 144″	g. 9″	j. 10″	m. 96″
b. 108″	e. 156″	h. 4 1/2″	k. 72″	n. 24″
c. 120″	f. 6″	i. 7 1/2″	l. 84″	o. 132″

3.
a. 1′	d. 4	g. 5 3/4 yds.	j. 23′ 3″	m. 2′ 10″
b. 3 yds. 2′	e. 38 yds.	h. 300 yds.	k. 19 yds. 2′	n. 6′ 9″
c. 9 yds. 1′	f. 11 5/12 yds.	i. 21′ 1″	l. 3′ 6″	

5.
a. 96 oz.	e. 3 lbs.	i. 16,000 lbs.	m. 6 s.t.	p. 8,960 lbs.
b. 128 oz.	f. 2 lbs.	j. 18,000 lbs.	n. 17.5 s.t.	q. 11,200 lbs.
c. 464 oz.	g. 3.5 lbs.	k. 26,000 lbs.	o. 20 s.t.	r. 15,680 lbs.
d. 520 oz.	h. 3.75 lbs.	l. 48,000 lbs.		

Unit 24 Perimeter

APPLICATION, Pages 148-149

1.
a. 5′ 10″	c. 194′ 4″	e. 5′ 5/12″	g. 18′ 6″	i. 20 yd.
b. 6.6″	d. 84 yd.	f. 7 yd 1′ 5 ¾″	h. 17 yd. 1′	

3.
a. 37.70′	c. 37.70″	e. 37.70 yd.	g. 15.71′
b. 18.85′	d. 37.70″	f. 56.55′	h. 15.71′

5.	P = 35′	13.	496′
7.	P = 44″	15.	25 7/8″
9.	C = 28.27″	17.	12′ 6″
11.	P = 59′	19.	50.2656″

Unit 25 Using the English System: Square Measurement

APPLICATION, Pages 155-156

1.
a. 864 sq. in.	c. 3,456 sq. in.	e. 432 sq. in.	g. 7,200 sq. in.
b. 2,736 sq. in.	d. 4,320 sq. in.	f. 2,304 sq. in.	h. 3,744 sq. in.

3.
a. 135 sq. ft.	c. 207 sq. ft.	e. 810 sq. ft.	g. 7,623 sq. ft.
b. 126 sq. ft.	d. 405 sq. ft.	f. 1,296 sq. ft.	h. 1,863 sq. ft.

5.
a. 2,592 sq. in.	d. 32,400 sq. in.	g. 2 sq. yd.	i. 8 sq. yd.
b. 12,960 sq. in.	e. 7,776 sq. in.	h. 4 sq. yd.	j. 4 1/2 sq. yd.
c. 19,440 sq. in.	f. 1 sq. yd.		

7.
a. 308 sq. in.	c. 38.05 sq. yd.	e. 5 sq. yd. 30 sq. in.
b. 5 sq. yd. 1 sq. ft. 6 sq. in.	d. 5 sq. ft. 47 sq. in.	f. 436 sq. yd.

Unit 26 Area of Squares, Rectangles, Triangles and Circles

APPLICATION, Pages 162-163

1. a. 100 sq. ft.
   b. 225 sq. ft.
   c. 2704 sq. ft.
   d. 1122 ¼ sq. in.
   e. 11 1/9 sq. ft.
   f. 30 ¼ sq. ft.
   g. 25 sq. yd.
   h. 32 1/9 sq. yd.

3. a. 252 sq. ft.
   b. 19.64 sq. ft.
   c. 63 sq. in.
   d. 144 sq. ft.
   e. 169 sq. ft.
   f. 6 sq. ft.
   g. 44.18 sq. ft.
   h. 96 sq. in.
   i. 20.25 sq. in.
   j. 79.2 sq. in.
   k. 95.03 sq. ft.

5. A = 143 sq. ft.
7. 13 1/3 or 14 sheets (Note: the ceiling is 8 ft. high.)
9. $562.50
11. 2445 sq. ft.

Unit 27 Area of Parallelograms and Trapezoids

APPLICATION, Pages 168-169

1. a. 32 sq. ft.
   b. 32 sq. yd.
   c. 81 1/4 sq. ft.
   d. 140.25 sq. ft.
   e. 85 sq. ft.
   f. 32 sq. in.
   g. 38.72 sq. in.
   h. 140.25 sq. in.
   i. 199 17/18 sq. ft.

3. A = 96 sq. ft.
5. 329.18 sq. ft. of sheathing
   272.25 sq. ft. or 2.7225 squares of shingles
7. 1250 sq. ft.

9. A = 17 13/18 sq. ft.
11. A = 402 1/9 sq. yd.
13. 5/9 lbs.
15. 25/144 lbs.

Unit 28 Area of Rectangular Solids

APPLICATION, Pages 172-173

1. a. 664 sq. ft.
   b. 804 sq. ft.
   c. 215.5 sq. ft.
   d. 42 4/9 sq. yd.

3. a. 864 sq. in.
   b. 486 sq. in.
   c. 1350 sq. in.
   d. 3456 sq. in.
   e. 6144 sq. in.
   f. 150 sq. ft.
   g. 600 sq. ft.
   h. 384 sq. ft.
   i. 1944 sq. ft.
   j. 2646 sq. ft.
   k. 181.50 sq. ft.
   l. 346.56 sq. ft.
   m. 496.86 sq. ft.
   n. 699.84 sq. ft.
   o. 922.56 sq. ft.

5. 63 sq. ft.
7. 242 2/9 sq. ft.
9. $572.00

11. A = 120 sq. ft.   wt. = 90 lbs.
13. 1176 sq. ft.
15. $72.00

Unit 29 Lateral Area and Total Area of Cylinders

APPLICATION, Pages 177-178

1. a. 100.53 sq. in.
   b. 43.98 sq. in.
   c. 100.53 sq. ft.
   d. 9.9484 sq. yd.
   e. 3.6652 sq. yd.
   f. 75.40 sq. ft.
   g. 31.42 sq. ft.
   h. 188.5 sq. ft.
   i. 62.43 sq. ft.
   j. 4.084 sq. yd.

3. A = 402.12 sq. in.
5. 63.56 sq. ft.
7. 735.13 sq. in.
9. 188.49 sq. ft. of insulation; $47.12
11. $23.33

Unit 30 Using the English System: Cubic Measurement

APPLICATION, Pages 183-184

1. a. 6912 cu. in.
   b. 15,552 cu. in.
   c. 13,824 cu. in.

   d. 34,560 cu. in.
   e. 148,608 cu. in.
   f. 14,688 cu. in.

   g. 3,888 cu. in.
   h. 9,504 cu. in.
   i. 10,800 cu. in.

   j. 4,752 cu. in.
   k. 8, 640 cu. in.

3. a. 108 cu. ft.
   b. 2,295 cu. ft.
   c. 2,916 cu. ft.

   d. 21,600 cu. ft.
   e. 2,667.6 cu. ft.
   f. 1417 1/2 cu. ft.

   g. 135 cu. ft.
   h. 2,052 cu. ft.
   i. 4,914 cu. ft.

   j. 56,160 cu. ft.
   k. 2,730.24 cu. ft.

5. a. 93,312 cu. in.
   b. 186,624 cu. in.
   c. 69,984 cu. in.

   d. 34,992 cu. in.
   e. 233,280 cu. in.
   f. 326,592 cu. in.

   g. 34,992 cu. in.
   h. 5,832 cu. in.
   i. 489,888 cu. in.

   j. 583,200 cu. in.
   k. 41,990.4 cu. in.

7. 22 cu. ft.
9. 5 cu. yd.
11. 1 cu. yd. 18 cu. ft.

Unit 31 Volume of Rectangular Solids and Cubes

APPLICATION, Pages 190-191

1. 17,600 cu. in.
3. 72 cu. ft.
5. 480 cu. yd.
7. Divide the volume in cubic inches by
   1,728 cu. in./cu. ft.

9. Divide the volume in cubic feet by
   27 cu. ft./cu. yd.
11. 3 15/27 cu. yd.
13. $192.59
15. 1 cu. ft.

Unit 32 Volume of Cylinders, Cones, Pyramids, and Spheres

APPLICATION, Pages 195-196

1. V = 402.12 cu. in.
3. V = 127.68 cu. ft.
5. The containers hold equal amounts.
7. V = 62.832 cu. ft.
9. V = 9,900 cu. in.
11. V = 137.5 cu. ft.
13. V = 209.44 cu. in.

15. V = 293.216 cu. in.
17. a. V = 5,575.29 cu. in.
    b. V = 65.45 cu. in.
    c. V = 14.1372 or 14.14 cu. in.
    d. V = 38.79 cu. in.
    e. V = .2209 cu. in.
    f. V = 137.259 or 137.26 cu. in.

SECTION 5 THE METRIC SYSTEM OF MEASUREMENT

Unit 33 Using the Metric System: Linear, Volume, and Weight Measurements

APPLICATION, Pages 200-201

1. a. 80 mm
   b. 120 cm
   c. 60 dm

   d. 310 dm
   e. 105.0 mm
   f. 84 cm

   g. 208 dm
   h. 880 cm
   i. 450 cm

   j. 550 cm
   k. 1 855.00 cm

3. a. 7 000 mm^3 b. 8 500 cm^3 c. .3 cm^3 d. 6.8 m^3 e. 38 m^3

5. 60 mm

7. 3.6 cm

Unit 34 Using the Metric System: Square and Cubic Measurement

APPLICATION, Pages 203-204

1. 76 800 mm^2
3. 450 cm^2
5. 2 dm^2

7. 357 m^2
9. 340 cm^2
11. 3 000 cm^3

SECTION 6 SQUARE ROOTS
Unit 35 Squaring and Square Roots

APPLICATION, Pages 208-209

1. a. 25   d. 121   g. 169   j. 1089   m. 15,625
    b. 81   e. 144   h. 196   k. 49   n. 90,000
    c. 100   f. 36   i. 484   l. 2704   o. .04

3. a. 5′   d. 12′   g. 28.28′   j. 7.2′
    b. 15′   e. 26″   h. 27.04′   k. 14.14′
    c. 24′   f. 72″   i. 27.22′   l. 50′

5. 22′

Unit 36 Hypotenuse

APPLICATION, Pages 213-215

1. a. 7.21′   d. 8.25′   g. 11.18′   i. 12.21′
    b. 8.49′   e. 12.65′   h. 12.53′   j. 11.40′
    c. 6.71′   f. 8.94′

3. 19.70′
5. 20′
7. 16.64′

SECTION 7 FUNDAMENTALS OF ALGEBRA
Unit 37 Introduction to Equations and Formulas

APPLICATION, Pages 219-220

1. a. $x + 6 = 18; x = 12$
    b. $x + 5 = 25; x = 20$
    c. $x + 4 = 32; x = 28$
    d. $15 - x = 7; x = 8$
    e. $12 - x = 4; x = 8$
    f. $8 - x = 2; x = 6$
    g. $5x = 25; x = 5$
    h. $6x = 18; x = 3$
    i. $4x = 24; x = 6$
    j. $\frac{x}{4} = 4; x = 16$
    k. $\frac{x}{5} = 6; x = 30$
    l. $\frac{x}{7} = 9; x = 63$

3. $I = E/R; I = .5$ amperes

5. a. $A = bh$ (A = area; b = base; h = height)
    b. $A = \pi r^2$ (π = 3.1416; r = radius)
    c. $V = LWH$
    d. $A = LW$ (L = length; W = width)
    e. $V = S^3$
    f. $V = 1/3\ Bh$

Unit 38 Variables and Constants

APPLICATION, Pages 223-224

1. A variable is a number whose value may change under different conditions.

3. A constant is a number in a formula that has only one value.

5. $A = \pi r^2$

7. The variables are W' and L' (width and length). The constant is 100.

9.
   a. variables P, L, W   constant 2, 2
   b. variables S, P   constant 4
   c. variables d, r   constant 2
   d. variables R, d   constant 2
   e. variables d, c   constant π
   f. variables c, d   constant π
   g. variables R, c   constant 2, π
   h. variables c, a, b   constant (none)

Unit 39 How to Transpose Equations

APPLICATION, Page 230

1.
   a. $x = 3$ d. $a = 2$ g. $b = 7$ i. $x = 7$
   b. $x = 6$ e. $a = 2$ h. $t = 6$ j. $a = 6$
   c. $x = 7$ f. $b = 4$

3.
   a. $x = 6$ d. $s = 6$ g. $g = 5$ i. $b = 6$
   b. $a = 1$ e. $s = 6$ h. $g = 3$ j. $c = 0$
   c. $b = 7$ f. $a = -2$

5.
   a. $x = 3$ d. $a = 2$ g. $x = 8$ i. $x = 6$
   b. $x = 2$ e. $a = 3$ h. $c = 2$ j. $h = 5$
   c. $a = 4$ f. $a = 7$

Unit 40 How to Determine Letter Value in Formulas

APPLICATION, Pages 233-235

1. 6 bd. ft.
3. $I = 14'$
5. $P = 20'$
7. $F = 32°$
9. $LH = 9'$
11. $I = 2$

13. $G = 5625'$
15. $R = 1.8462$
17. $I = 12'$
19.
   a. $I = 1\ 1/2'$
   b. $w = 11'$
   c. $A = 143$ sq. ft.

Unit 41 How to Solve Equations by Square Roots

APPLICATION, Pages 239-241

1.
   a. .87   d. .50   g. .79   j. .94   m. .89   p. .67
   b. .82   e. .52   h. .75   k. .76   n. .63   q. .68
   c. .58   f. .60   i. .56   l. .52   o. .75   r. .63

3. $d = 13.54''$
5. $r = 4.9$ ft.

SECTION 8 FUNDAMENTALS OF GEOMETRY
Unit 42 Circles and Polygons

APPLICATION, Page 250

1. Eccentric circles are two or more circles having different centers. Concentric circles are two or more circles having a common center.

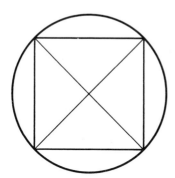

3. Square

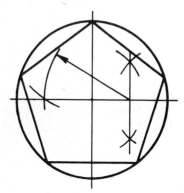

5. Pentagon

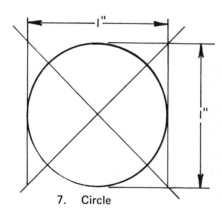

7. Circle

9. r = 3.16″

Unit 43 Angles

APPLICATION, Pages 256-257

1.
a. 300′	c. 480′	e. 1320′	g. 4500′	i. 442′
b. 420′	d. 1200′	f. 780′	h. 330′	j. 1928′

3.
a. 1740″	c. 3000″	e. 2100″	g. 427″	i. 2257″
b. 1140″	d. 3120″	f. 907″	h. 3510″	

5. a. 90°   c. 30°   e. 75°   g. 20°   i. 17°

b. 60°   d. 15°   f. 35°   h. 25°   j. 40°

7. a. 5° 13′ 8″   b. 4° 45′ 45″   c. 3° 49′ 50″   d. 2° 20′ 59″

9. a. 4° 5′ 38″   b. 13° 39′ 4″   c. 7° 36′ 44″   d. 7° 11′ 9″

SECTION 9 FUNDAMENTALS OF TRIGONOMETRY

Unit 44 Right Angle Trigonometry

APPLICATION, Pages 264-265

1. Sine (sin) is the ratio of the side opposite to the hypotenuse.

3. Cosine (cos) is the ratio of the side adjacent to the hypotenuse.

5.
a. .50000	d. .70711	g. .90631	i. 1.0000
b. .34202	e. .93969	h. .98769	j. .99985
c. .32557	f. .81915		

7.
a. .96126	d. .84805	g. .48226	i. .75851
b. .64279	e. .88295	h. .80558	j. .52498
c. .30902	f. .73728		

Unit 45 The Sine Function

APPLICATION, Pages 271-272

1.
a. .5000	d. .6667	g. .6667	i. .9167
b. .6667	e. .6649	h. .6429	j. .857
c. .7037	f. .6000		

3.
a. .6667	d. .7000	g. .8407	i. .7211
b. .5714	e. .9091	h. .8175	j. .7862
c. .6667	f. .8190		

5. $\angle B = 60°$

7. rise = 6.25′ or 6 ¼′

Unit 46 The Cosine Function

APPLICATION, Pages 279-280

1.
a. 36° 50′	e. 41° 20′	i. 41° 20′	m. 51°
b. 41° 20′	f. 39° 00′	j. 43° 20′	n. 51° 40′
c. 48° 10′	g. 43° 20′	k. 32° 40′	
d. 44° 20′	h. 44° 50′	l. 38°	

3. $\angle A = 60°$

5. $\angle A$ = approx. 29° 00″

Unit 47 The Tangent Function

APPLICATION, Pages 286-287

1.
a. .10510	e. .12278	i. .46631	m. 1.0000
b. .23087	f. .24933	j. .57735	n. 1.1918
c. .21256	g. .32492	k. .70021	o. 1.4282
d. .15838	h. .36397	l. .83910	p. 1.7321

3. a. 33° 40'
    b. 36° 50'
    c. 37° 50'

    d. 52° 10'
    e. 51° 20'
    f. 26° 30'

    g. 35° 20'
    h. 35° 30'
    i. 57° 50'

    j. 51° 30'
    k. 40° 50'
    l. 39° 30'

5. ∠A = 37° 30'

7. run = 16.4 ft.

Unit 48 Interpolation

APPLICATION, Page 290

1. a. 9° 45'
    b. 12° 5'
    c. 15° 45'

    d. 17° 55'
    e. 20° 5'
    f. 22° 25'

    g. 20° 45'
    h. 21° 15'
    i. 22° 42'

    j. 24° 20'
    k. 25° 46'
    l. 16° 35'

3. a. 64° 42'
    b. 58° 36'
    c. 19° 40'

    d. 8° 10'
    e. 57° 55'
    f. 55° 25'

    g. 8° 5'
    h. 11° 19'
    i. 24° 39'

    j. 18° 55'
    k. 21° 5'
    l. 22° 5'